ON TRANSHUMANISM

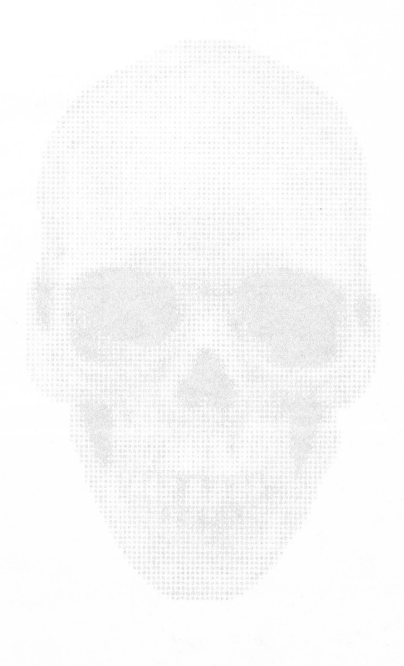

ON TRANS-
HUMANISM

Stefan Lorenz Sorgner
TRANSLATED BY SPENCER HAWKINS

The Pennsylvania State University Press
University Park, Pennsylvania

Library of Congress Cataloging-in-Publication Data

Names: Sorgner, Stefan Lorenz, author. | Hawkins, Spencer, translator.
Title: On trans-humanism / Stefan Lorenz Sorgner ; translated by Spencer Hawkins.
Other titles: Transhumanismus. English | On transhumanism
Description: University Park, Pennsylvania : The Pennsylvania State University Press, [2020]
    | Translation of Transhumanismus : "die gefährlichste Idee der Welt"!?. | Includes
    bibliographical references and index.
Summary: "Examines widespread myths about transhumanism and explores the
    most pressing ethical issues in the debate over technologically assisted human
    enhancement"—Provided by publisher.
Identifiers: LCCN 2020036961 | ISBN 9780271087931 (paperback)
Subjects: LCSH: Transhumanism.
Classification: LCC B842.5 .S6713 2020 | DDC 144—dc23
LC record available at https://lccn.loc.gov/2020036961

This volume includes edited and expanded versions of the following articles, which have
been published elsewhere: "Nietzsche, der Übermensch und Transhumanismus," in
Enhancement und Genetik, edited by N. Knoepffler and J. Savulescu (Freiburg i. B.: Alber
Verlag, 2009), 127–44; "Enhancement von Emotionen und Moral: Eine Landkarte der
Verbesserungsdebatte," in *Medizin, Moral und Gefühl: Emotionen im ethischen Diskurs;
Jahrbuch Ethik in der Klinik (JEK)*, edited by A. Frewer, F. Bruns, and W. Rascher, 5th ed.
(Würzburg: K&N, 2012), 157–76; "Posthumane leben besser: Ist der Transhumanismus
die gefährlichste Idee der Welt?," *Aufklärung und Kritik* 19, no. 4 (2012): 160–73; and
"Stammbäume des Meta-, Post- und Transhumanismus" and "Plädoyer für einen schwachen
Transhumanismus," in "Schwerpunkt Transhumanismus," edited by S. L. Sorgner, special
issue, *Aufklärung und Kritik* 22, no. 3 (2015): 4–27 and 273–290.
Translation and "Translator's Introduction: Transhumanism in Translation" by Spencer
Hawkins.

The Pennsylvania State University Press is a member of the Association of University
Presses.

It is the policy of The Pennsylvania State University Press to use acid-free paper.
Publications on uncoated stock satisfy the minimum requirements of American National
Standard for Information Sciences—Permanence of Paper for Printed Library Material,
ANSI Z39.48–1992.

CONTENTS

ACKNOWLEDGMENTS

I would like to thank Pascal Kreuder, Markus Peuckert, Robert Ranisch, Alina Sorgner, and Sarah Uhrig for reading this manuscript and offering helpful feedback. I am especially grateful to the Alber Verlag's editor-in-chief, Lukas Trabert, with whom I have enjoyed constructive, friendly, and successful collaboration over many years. I am especially indebted to Manuel Herder, who enabled this work's original publication with Herder Verlag.

TRANSLATOR'S INTRODUCTION

Transhumanism in Translation

This introduction situates Stefan Lorenz Sorgner's *On Transhumanism* within the thematic context of technological innovation, but also within the situational context of the tools and techniques available for translating German philosophy into English. In this book, Sorgner speaks highly of controversial biotechnologies, like xenotransplantation, which would replace, for instance, damaged human lungs with healthy lungs from a pig bred for that purpose. Like an organ transplant in words, a translation introduces foreign ideas into a new cultural body, which could easily reject those ideas just for being foreign. Among other goals, this introduction intends to provide a kind of antirejection drug to offset effects of this book that might lead North American readers to put it down too soon.

THIS TRANSHUMANIST MOMENT

What unites transhumanists is an enthusiasm for research and development into life-changing technologies. They want more technology legalized and made widely available faster, and they come with varied motivations, including curiosity, conviction, medical necessity, and financial interest. In transhumanist visions of the future, unprecedented technologies will release human lives from bodily constraints that medicine has thus far regarded as inevitable. Scientists are now in a position to translate the human genetic code out of the messy handwriting of nature into the exquisite calligraphy of our wildest dreams. More relaxed regulations on new biotechnology (especially gene modification) will open the door to life in the superlative: superior health, supreme cognitive functioning, unheard of athletic potential, and super-well-being. *On Transhumanism* weighs in on the ethical questions around the pursuit of technologically assisted self-enhancement.

It is a defining feature of technology that it exists to serve human wishes. What is so controversial then about saying that it is better to have more sooner? The following examples give a sense of the controversial flair of transhumanist enterprises. Since the late 1970s, at the cryogenics facility in Scottsdale, Arizona, Alcor Life Extension Foundation has been "extending medical care past when the doctors give up on you," as one member describes it (Alcor Cryonics 2015). An Alcor member becomes a patient immediately upon clinical death, at which point "cryoprotectant concentration" is circulated through patients' major arteries. They are then laid in liquid nitrogen-cooled chambers where they are preserved indefinitely.

Even though Alcor has not attempted to revive a patient yet, "it is assumed that the cryopreservation process will someday be reversible" (Alcor Procedures, n.d.). Cryogenics falls within the kind of experimental medicine not covered under health insurance policies, and therefore members typically secure payment to Alcor by naming it as the beneficiary of life insurance policies, taken out in the amount of the $200,000 full-body preservation fee—or $80,000 for the head alone. Members can also pay to have pets preserved for up to $30,000. One Alcor member cancelled her cell phone plan to pay the $600 yearly membership fee and explained to a reporter: "I'm sorry, but I'm just not that excited about phones. I'm excited about teleportation devices or my own personal spaceship. I want to see the future" (CNBC 2016). Of course, Alcor cannot promise that the procedure is reversible, and other futurists, like Michio Kaku and this book's author, are not optimistic that cryonics will work.

Like so many health care procedures, self-optimization tends to be expensive, and it tends to break skin. From late 2016 to early 2019 (at which point the Food and Drug Administration intervened), a San Francisco-based company called Ambrosia was injecting blood plasma from donors under twenty-five, purchased from blood banks, to older clients for $8,000 per liter.[1] Ambrosia hoped to replicate the antiaging effects of this procedure (known as parabiosis) that improved the appearance of fur and internal organs in mouse trials. Although there have been no human test trials, one Ambrosia client, a sixty-two-year-old mathematician with injuries from a motorcycle accident, reported that his sleep improved significantly after a year of infusions (Carville 2019).

An example of a cheaper, less speculative "biohack" is the microchipping of people. Microchipping is already the favored way of tracking house pets, and in 2015, Biohax International began implanting microchips in the webbing between people's thumb and index finger for only $180 per chip. Within Sweden, where the company operates, thousands of residents rely on their implants in place of credit cards, train tickets, and passports in everyday transactions.[2] E-tickets and online payments are similarly convenient and do not require chip implants (which were, after all, first implemented as tracking devices), but in the world's most paperless society, where fewer than 1% of transactions involve cash, the motivation to streamline transactions is higher than elsewhere (Alderman 2018).

This brief overview gives a sense of the sci-fi appeals of transhumanist enhancements on human existence: from mere convenience to enhanced vitality to immortality. Although the effectiveness of some of these technologies remains speculative, what matters to transhumanists is the relaxation of laws around innovative biotech. Perhaps more so than in other luxury industries, the clientele is mostly male.[3] Transhumanists frame their visions as inclusive of all humans, though, and Sorgner insists that transhumanist goals could benefit nonhuman animals and artificially intelligent machines as well. The US Transhumanist Party thus demands research with the *intention* of improving life "should be rendered fully lawful and their products should be made fully available to the public, as long as no individual is physically harmed without that individual's consent or defrauded by misrepresentation of the effects of a possible treatment or substance" (US Transhumanist Party / Transhuman Party—Official Website, n.d.). By expressing the wish for self-optimization in familiar political language, transhumanism calls for its place in the contemporary policy discussion. Academic work on transhumanism, like the present book, strikes a cooler tone than the movement's political and economic spokespeople, but their shared enthusiasms make these discourses difficult to disentangle.

THE PRESENT BOOK IN CONTEXT

The English edition of this book can be read as a kind of reverse translation in the sense that the transhumanist movement has already had a more visible following in the Anglophone world than in Europe, as Sorgner points out, and this book originally addressed a German-speaking audience. The US Transhumanist Party, for instance, was the first political party dedicated to the transhumanist movement in the world, and—although American transhumanists come in all political stripes—the party's deregulatory platform appeals to the libertarian contingent so prevalent in the United States. Within Europe, Swedes show the most eager embrace of transhumanism. Swedish society reportedly exhibits fewer fears than its neighbors about data privacy, which some believe explains why they have been quicker to implement transhumanist ideas, like microchipping. Another possible reason why transhumanism commands more respect in Sweden than in other European countries is that Swedes regard their tech sector as foundational to their prosperous economy (Petersén 2018). Whatever the reason, Swedish law is what transhumanists would call "bioliberal" when it comes to legalizing elective enhancement technologies. Since Sweden also provides government-sponsored health care to all its citizens, its laws most closely match the stance of this book's German-born author, who brings European political sensibilities to the transhumanist movement—otherwise dominated by the voices of American tech CEOs.

Sorgner is currently a philosophy professor at John Cabot University in Rome. His centrality in the transhumanist discourse is unmistakable: he is a prolific author, editor of a journal and book series on transhumanism, sought-after as a public speaker, and trained as a Nietzsche scholar. He completed his PhD with Gianni Vattimo, author of over one hundred philosophy books, who himself argues, drawing on Heidegger and Nietzsche, for a "weak" concept of Being that he finds compatible with both Christianity and nihilism. Sorgner is an equally idiosyncratic philosopher. The present book positions transhumanism in unexpected ways philosophically, institutionally, and politically. It argues that Nietzsche's ethics of self-overcoming, his ontology of power, and his quasi-Lamarckian evolutionary views can (and should) be read as supportive of transhumanism, and that a weak transhumanism, or metahumanism, rejects the premises of humanism in ways compatible with minority rights discourses and leftist posthumanism. The author describes transhumanism as politically unspecific enough for its adherents to combine it with political views as various as libertarianism, classical liberalism, and social democracy.

In 2018, Sorgner spoke on a panel with other Nietzsche scholars, as part of the radio-broadcast Phil.Cologne public debate series. After pointing out that enhancement technologies (vaccines, Viagra, smartphones—our omnipresent sixth sense) already structure our everyday life, Sorgner described a new enhancement on the horizon: implants to measure blood sugar, which could detect oncoming insulin shocks before a person suffered symptoms. The moderator expressed perplexity at Sorgner's great enthusiasm for the implant: "You would have to be awfully *worried* about yourself to be so excited about such a device. Are you so worried about yourself?"[4] Sorgner replied that he considered blood sugar levels a generally reasonable concern. The question with any new technology, though, is whether it raises more worries than it addresses. In the case of real-time health monitoring tools, one concern (that Sorgner raises below) would be that insurance companies could access the data, discover symptoms before the client even notices them, and disqualify clients for care on the basis of preexisting conditions.[5]

In its adoration for tech, transhumanism can look like the inversion of the antiresearch platform of America's religious right, and transhumanism occasionally dovetails with leftist positions. This book argues for the ethical legitimacy of a new reproductive technology legalized in Britain in 2015 that can produce a child with sex cells from three biological parents. That technology has found some resonance with cyber-feminist leader Donna Haraway, who sees three-parent fertilization as an alternative to patriarchal heterosexual coupledom (Haraway 2016, 8, 138). Yet even when such points of overlap occur, most academics see transhumanism's *premises* as glorifying idealized humanity over everything and everyone else (including humans deemed inferior). Since the book primarily defends transhumanism against its German-speaking critics, I will introduce some Anglophone critical positions below in greater detail. But first I want to introduce transhumanism's most prominent supporters—mostly based in California—some of whom are briefly mentioned in this book.

OF UNIVERSITY PROFESSORS AND TECHNOLOGY EXECUTIVES

In the fourth chapter, Sorgner summarizes Nietzsche's view that powerful individuals shape our perceptions of reality, even our perception of power itself: "The content of the concepts of power is always bound to the perspective of the powerful." The media portrayal of transhumanism in the Anglophone world is highly attuned to this Nietzschean insight and focuses on the wealthiest transhumanists: technology executives, like Elon Musk and Peter Thiel. I will therefore devote this section of the introduction to supplementing the book's focus on transhumanist ideas by introducing the social positions of transhumanism's most prominent academic critics and its most influential spokespeople, beginning with the latter.

This book introduces Nick Bostrom and Natasha Vita-More as a Swedish-born Oxford professor and an American artist; they are also cofounder and executive director of Humanity Plus, respectively. Humanity Plus is a think tank whose mission is "to deeply influence a new generation of thinkers who dare to envision humanity's next steps" (Humanity+, n.d.). Their goal of increasing public acceptance of biotechnology has attracted an international base as well as some questionable American donors. Humanity Plus recently made headlines for having benefitted from notorious billionaire eugenicist and convicted sex trafficker Jeffrey Epstein's $20,000 donation in 2011 (Stewart, Goldstein, and Silver-Greenberg 2019).

Bostrom and Vita-More make attention-getting statements in keeping with their role as spokespeople for the transhumanist movement. Natasha Vita-More for instance has raised the possibility of holding a "Super Olympics" with biologically enhanced athletes.[6] Just picturing that image could reframe the doping debate for some sports fans: anti-doping regulations are standing between investors and their dreams of sponsoring supersports for bioenhanced titan-athletes whose feats will be more entertaining than we can even imagine today.[7] Bostrom, a philosophy professor at Oxford University, has also worked hard to reframe the discourse; he offers a memorable analogy between the mainstream global tendency to accept death as inevitable and the mores of a fictional, backward society that practices human sacrifice and tells children to accept "the sacrifices . . . as a fact of life" (Bostrom 2005c). Through such images and analogies, transhumanist intellectuals enter a cultural war by raising an optimistic bulwark against dystopian films that portray enhancement technologies as primarily oppressive or exploitive, such as *Gattaca* (1997) or *Get Out* (2017).

Tech journalists tend not to report heavily on these transhumanist intellectuals. They instead follow the money by asking who stands to gain most directly by capitalizing on the enhancement technology market. The answer is clearly those with capital invested in biotech industries. For decades now, one of the most prominent transhumanist entrepreneurs has been Ray Kurzweil, director of engineering at Google. After revolutionizing keyboard synthesizers in the 1980s, he turned to biotechnology in the 1990s. He is most famous today for predicting that a singularity event will occur before the year 2050, wherein humans will no longer rely on bodies to process information because they can interact directly over a data cloud (Kurzweil 2005). When a Google executive promises mind-uploading within our lifetimes, a signal to investors cannot help but slip into the prediction: Google technology is about to change the world again, and now is the time to invest *more capital* in Google or else miss out on massive profits.

Other tech CEOs advertise their transhumanism in ways that sound designed to shock. PayPal founder, biotech investor, and outspoken Trump supporter Peter Thiel has expressed interest in receiving transfusions of young people's blood as a way of fighting off death, like what Ambrosia was offering (Kosoff 2016). Tesla's Elon Musk argues that autonomous vehicles will soon replace all jobs involving the driving of automobiles, but he comes prepared with a risky, dystopian-sounding solution: turning unskilled workers into highly efficient cyborgs. In his words: "Some high bandwidth interface to the brain will be something that helps achieve a symbiosis between human and machine intelligence and maybe solves the control problem and the usefulness problem" (Clark 2017). Such workplace innovations sound especially risky coming from someone whose labor practices around health and safety at Tesla are already under close scrutiny (O'Kane 2018). In his 2018 book, *Schöner Neuer Mensch* (Brave new human), Sorgner argues that prominent transhumanists, such as Elon Musk, make absurd statements—like claiming that we currently live in a computer simulation à la *The Matrix*—because good or bad media attention enhances their brand recognition.

Tech entrepreneurs may seem especially prone to overblown statements, and those statements may dominate the public conversation excessively, but it makes sense to pay attention to them. After all, they are motivated to be well informed about the potential applications of new technology and thus are poised to present provocative bioethical arguments. Dan Faggella, for instance, is CEO and founder of TechEmergence, a marketing research company dedicated to promoting artificial intelligence technologies. Despite his investment in the industry, even Faggella warns that two dangerous extremes of bioenhanced humans will emerge: the "lotus-eaters," who will use AI and bioenhancement to experience escapist pleasures, and the "power-eaters," who will use simulations and self-enhancements to train harder, sleep less, feel fewer distracting emotions, and accomplish more than their rivals: "In the coming century, almost all economic competition, political competition, and war will ultimately be a proxy for obtaining this pinnacle of technological control and power" (Faggella 2018). Faggella's warnings make historical sense when we consider that the internet had its first instantiation in ARPANET, the US military network designed to enhance geopolitical control in Southeast Asia, and that the United States primarily conducts its twenty-first-century wars in front of computer screens (Levine 2018, 13–35). Technology empowers the most powerful most of all. This historical context must be addressed for a transhumanist ethos to be persuasive.

Persuasive ideas for a just and pluralistic posthuman future do sometimes come from biotech industry leaders, such as "transgender transhumanist" Martine Rothblatt, former outer-space-domain lawyer, founder of SiriusXM radio, and current CEO of United Therapeutics, an experimental pharmaceutical company.[8] Throughout the book, Sorgner pleads that laws should be less restrictive against research on technology that could transform human potential. Rothblatt frames transhumanism as a matter of minority rights in her blog, books, interviews, and articles. Sorgner makes the same move, for instance, when he discusses three-parent fertilization. This argument for transhumanist reform shifts the focus away from individual achievement toward forms of collective solidarity.

Transgender rights may even be the most provocative "social justice" case for transhumanism. Some transgender people identify as neither "he" nor "she," and they suffer from discrimination wherever binary gender identifications are legally required: when they go to the restroom, complete an application for work, or apply for government or medical services. In the 1960s, medical articles began to report that some patients regarded their need for sex-change surgery as a matter of life or death. For many trans people, a socially ostracized life full of special medical needs is still preferable to the debilitating depression they suffer before their surgery (Meyerowitz 1980). The struggles for legal recognition by those who experience gender-related body dysmorphia makes an apt analogy for those transhumanists for whom laws against mental or physical enhancement (by surgery or medication) stand in the way of biotechnological optimization. When Rothblatt says in a TED talk, "There are seven billion people on this earth, and there are seven billion unique ways to express one's gender," she calls into question the law's codification of gender (Rothblatt 2015). Rothblatt's transhumanism (which Sorgner finds compelling) opposes the state's right to endorse some biotechnologies as medically necessary[9] and to reject others on grounds of being excessively self-interested. Writers who rely on amphetamines like Adderall face a stigma similar to transgender people: because they were medicated, their success is therefore regarded as less "real." While it is a privilege to have access to medications and to sex reassignment surgery, and the government should probably still regulate new technologies, transhumanism may have its ethical center in rejecting notions of "realness" and "naturalness" that stigmatize getting high-tech help.

Let us briefly survey transhumanism's various critics. After all, this book takes its polemical tone in response to them. The book's subtitle refers to a special issue of *Foreign Policy* magazine where each contributor discusses one of "The World's Most Dangerous Ideas." Francis Fukuyama, the center-right political scientist and regular guest author at *Foreign Policy* who once codeveloped the Reagan Doctrine but more recently came out against the Iraq War, selects transhumanism as the "most dangerous" contemporary idea because its deregulatory impulse could change our bodies and societies irreversibly before we know it: "The seeming reasonableness of the project, particularly when considered in small increments, is part of its danger. Society is unlikely to fall suddenly under the spell of the transhumanist worldview. But it is very possible that we will nibble at biotechnology's tempting offerings without realizing that they come at a frightful moral cost" (Fukuyama 2004).

In 2002 Fukuyama had already published an entire book warning against the risks of bioengineering (Fukuyama 2002), but in the very short 2004 *Foreign Policy* piece cited above, Fukuyama condenses his bioconservative argument against genetic engineering in the name of preserving the "essence" of the human "at the heart of political liberalism." If we do not guard against the transhumanists' "genetic bulldozers and psychotropic shopping malls," the risks would be the neglect of those "left behind" and the possibility that posthumans would be so morally different that they may not even be worthy of human rights. Beyond the political dimension, Fukuyama warns that we cannot anticipate the biological risks for humanity's survival if genetic modifications were widely adopted.

Sorgner opposes this view with a bioliberal stance he articulates at the end of the first chapter: "I believe that constant self-overcoming is central to promoting my own quality of life. I also consider scientific research, especially in biotechnology, extremely important and advocate for greater sponsorship of those research fields. I consider the availability of anesthetics, vaccinations, and antibiotics important achievements. I hope that further achievements will follow to address important challenges. This stance can be parsed as a weak form of transhumanism." Sorgner calls his stance "weak" transhumanism because it leaves the choice to adopt emergent biotechnologies up to individuals. Sorgner thinks that most other transhumanists basically concur: "Transhumanists embrace the liberal-democratic order as foundational and thus attach great importance to the norms of freedom and equality." At the same time, he makes it clear that a few outspoken "strong" transhumanists argue publicly that genetic engineering and antiaging technologies are morally imperative and that some of these not-so-liberal, "strong" transhumanists do lobby for the use of new biotechnologies to be *required* universally.

Reviewers of the German version of the book reflect the extent of the controversy. One German bioethicist, Konrad Ott, points out that Sorgner's reasons for believing that humanity will evolve into a new species through biotechnology "ultimately remain unexplained." And another German bioethicist, Marcus Knaup, finds the book intellectually unsatisfying and ethically disquieting: "Considering that it advocates liberal eugenics as a self-evident good, draws a grotesque image of humanity, and rejects the core of the ethics of reason by attacking the notion of human dignity, it is indeed a dangerous book, which has nothing to do with serious philosophy" (Knaup 2017, 35). Knaup dismisses Sorgner's arguments by calling on humanistic principles (Knaup is also a Roman Catholic theologian). Like Fukuyama, Knaup is appalled at transhumanism's rejection of traditional notions of "the human" in favor of a flux tethered only to the tech market. According to such worries, everything we associate with being human—war and social justice, medicine and pleasures, our family and social life, access to information, whatever we call culture —may be transformed irreversibly, not necessarily for the better, at the whim of a few entrepreneurs whose motivation is short-term profit.

Where humanists fear transhumanism for rejecting the image of humanity they find indispensable to a meaningful life, the movement is just as controversial among the posthumanists, who generally deny that transhumanism goes "beyond" humanism. Posthumanist philosopher Cary Wolf calls transhumanism "an intensification of humanism" for its focus on a generalized human experience to be set apart from the rest of nature and enhanced (Wolfe 2010, xv). Pramod Nayar explains further that transhumanism does not share posthumanism's skeptical insight because the former fails to see "the human as a construct enmeshed in other forms of life" and instead insists "that *there is a distinctive entity identifiable as the 'human'*" (Nayar 2014, 6–7). Nayar thus criticizes transhumanism for overlooking complexity at every turn: it is "techno-deterministic, and techno-utopian" in that it sees its goals as being "achieved almost exclusively through technology." Furthermore, transhumanism implies body-mind dualism in that it "relies on human rationality as a key marker of 'personhood' and individual identity, and sees the body as limiting the scope of the mind." Sorgner discusses these accusations over the course of the book, mostly by distancing himself from this or that transhumanist who does indeed fall directly into one of the simplistic claims that Nayar describes and insists that none of these *particular* views are *definitive* of transhumanism.

Why is complicity in humanism such a damning accusation? After all, humanism is linked to foundational modern ideals, like human rights and humanistic education. During the rise of colonialism, however, the exclusiveness of humanism was in full view as it was used to justify horrific violence against colonized people by linking humanity to the specifics of European upbringing and, when convenient, making pale skin a qualification for basic human rights. In the wake of the Holocaust, global thought leaders began calling humanism into question. Inspired by developments in feminism, decolonization, and civil rights, twentieth-century thinkers from Franz Fanon to Donna Haraway unmasked the implicit imperialism, white supremacism, speciesism, and misogynistic biologism of humanism by showing that there has always been more to belonging to humanity than possessing a set of biological traits. *Homo sapiens* are always already more than human owing to our deep enmeshment with both the natural and the technological world. Sorgner cites this theoretical overthrow of classical humanism under the name "critical posthumanism." And such critical rethinking of the human is currently redefining "humanistic" academic disciplines around the globe. Transhumanism too counters humanistic axioms, but by different means: with an aspirational program claiming that humans are in fact still all too human and that our humanity is holding us back from an unknown potential.

Because transhumanism advocates changes in the laws, behaviors, and attitudes around biotechnology, its adherents aspire to a better future, whereas "posthumanists are indifferent to the concept of progress," as Sorgner puts it in his third chapter. The tide-that-lifts-all-ships scope of transhumanism differentiates it from other aspirational alternatives to classical humanism, such as Afrofuturism, an artistic and intellectual movement, which performer Janelle Monáe defines as "us, Black people, imagining ourselves in the future . . . as magical as we want to be" (Zhou 2018). Spokespeople for specific minority groups often feel that their own group could be left stranded by technological process, a view well encapsulated in Gil Scott-Heron's poem: "I think I'll sen' these doctor bills, Airmail special, to Whitey on the moon" (Scott-Heron 1970).

If transhumanism is an enthusiastic response to progress thus far, Afrofuturism is a response to "the digital divide, a phrase that has been used to describe gaps in technological access that fall along lines of race, gender, region, and ability but has mostly become a code word for the tech inequities that exist between blacks and whites" (Nelson 2002). As scholar of Afrofuturism Alondra Nelson points out, "Blackness gets constructed as always oppositional to technologically driven chronicles of progress." Meanwhile, prominent futurists, like Timothy Leary and Allucquère Rosanne Stone, are "seemingly working in tandem with corporate advertisers." Some transhumanists, like Sorgner, claim that transhumanism is not meant to benefit only technocratic elites but is for everyone who desires novelty through biotechnology. Sorgner is aware that the most common sources of resistance to transhumanist visions are doubts—based on history and experience—that the benefits of new biotechnology could ever be distributed fairly.

Despite the ferocity of the debates, staking a position may have a minimal or even a reverse effect on policy. Leading artificial intelligence researchers like Yann LeCun warn that "AI winters," periods of stagnation in research and development, result from too much hype around what new technologies can offer (Marcus 2013). The public already finds achievements in AI underwhelming when the state of technology trails too far behind popular sci-fi scenarios, and a similar risk could accompany transhumanism's efforts to attract interest in biotech through promises of unprecedented new experiences. The best parts of the present book thus do not hype the technologies on the horizon, nor promise wonders, but remind us that further research and more bioliberal laws are still necessary to discover the most life-changing technologies.

Healthcare is already distributed extremely unequally in much of the world; self-optimizing technologies would presumably go to those who could afford to buy a competitive edge. How persuaded you are by Sorgner's response to that argument might be a Rorschach test for your views on technocracy in the present. Sorgner cherishes "liberalism" as the predominant political model that would achieve a just regulation of new biotechnologies. If you rate the successes of liberal regulation of technology well so far, then you may agree with him. If you are concerned about inadequacies in the distribution of health care today (in much of the United States, for instance), then you may be more skeptical. Although Sorgner does not endorse outright libertarianism, he opposes "patriarchal" states that ban, restrict, or criminalize the research, application, and marketing of self-enhancement technologies that would benefit individuals who wish to use them. The book makes gestures, however, to quell readers' fears that, in the posthuman future, the gene-rich will have a new form of capital to lord over the gene-poor, so that the human capital would sink even more for those who cannot afford enhancements (of strength, intelligence, perhaps advantageous forms of emotional coldness) in a competitive economy.

The present book's third chapter argues for a rapprochement between academic posthumanism and pragmatic transhumanism. Sorgner calls this middle ground metahumanism, which "strives to mediate among the most diverse philosophical discourses in the interest of letting the appropriate meaning of relationality, perspective, and radical plurality emerge." He argues that metahumanities would acknowledge the need for technologically mediated progress while also engaging in theoretical debates about the place of the human within the natural world. According to Sorgner, liberal laws on biotechnology are universally desirable on the grounds that liberalism generally is meant to account for the flourishing of all citizens. What the chapter does not discuss is the range of harms that might arise from enhanced humans' new potential. By contrast, when founding theorist of liberalism John Stuart Mill discusses why citizens must be legally entitled to potentially self-harming freedoms, for instance, to consume alcohol, he also discusses why laws must limit the freedom of drunk people to become nuisances (Mill 1859, 181). That is one problem for future work. But liberal theory has always born a sinister problem at its core: no matter how antipaternalistic liberal laws are, the histories of liberal nations notoriously thrive off of the exploitation of "barbarian" societies—a line of action that Mill endorses in the same essay. The most widespread current enhancements, like smartphones, already depend on cruel and environmentally unsustainable labor practices abroad. To stage a dialogue between transhumanism and the rest of academia would require the "metahumanists" to show as much curiosity about past and ongoing abuses in healthcare and biotechnology as they show about possible bright futures.

TRANSLATING TRANSHUMANISM

Translating from German (a language with around 4.2 million words) into English (a language with around 100 million words) yields abundant opportunities to transform and enhance a text (Jones and Tschirner 2015). In my research on translation, I have argued that one of the most important philosophical tasks of a translator is to look for distinctions that may not exist in a source language, and which the author thus downplays, and to introduce those nuances into the target text (*die Technik*, for instance, can mean "technique" or "technology"—as I discuss below). "Differential translation," as I call it, exposes the reader to the mental flexibility (or terminological conflations) that foreign languages offer (Hawkins 2017). Such translations empower readers to make an informed judgment as to whether they agree with the wisdom implicit in a source language, which sometimes combines concepts that their own language would distinguish, or whether they find the use of these words as terms imprecise. For this empowerment of the reader to work, translators must mark these moments of creativity in introductions, commentaries, or brackets so that readers can trace the use of whichever terms are transformed in the translation. Below I will discuss six words that required creativity in this translation: *er, bejahen, fördern, Technik, Technologie, Anthropologie*, and *Möglichkeit*.

Er simply means "he," but the German language contains an instance of grammatical sexism not present in English; it uses *er* in places where "he" would stand out as gendered in English. Furthermore, gender stands to be completely rethought in the transhuman future drafted in this book. At one point in the translation, I mark this transhumanist gender neutrality by turning the grammatical difference between English and German gender specificity into an opportunity to indicate that future humans will not conform to gender binaries: "Who is the posthuman? What qualities does he, she, they, or it have?" (*Wer ist der Posthumane? Welche Eigenschaften hat er?*)

Likewise, *bejahen* simply means "affirm," but in this book affirmation means support for ontologically varied objects: both for technology and for the values of daring, speed, and innovation. A variety of objects implies varied acts of support, which I represented with various words—including "defend," "embrace," "champion," "argue for," and "tout." This range of words was necessary to convey the transhumanist's multifarious enthusiasm for biotech research, development, and implementation. *Bejahen* defends the rhetorical citadel overlooking the crossroads between improved technology and enhanced lives. Freighting these words with ambiguity reinforces the book's main dare: legalize technology.

Standard English translations of *fördern* include "promote" and "support," but I sometimes translate it with "foster" or "enhance" when the context relates to the aims of biotech research (happiness, intelligence, health, etc.), as opposed to describing supporting the research itself. The argument describes a two-part dynamic process where *fördern* is the motor on both sides: if we support new technology, new technology will enhance us. "Technology can already promote (*fördern*) greater diversification of the means of human reproduction." In other words, the technology is already there to make reproduction serve humanity better than it currently does. But new technology still must be implemented in order to foster human thriving: "The traits and capacities that are especially relevant for fostering a good life (*eine Förderung des guten Lebens*) are emotional, psychological, and intellectual capacities along with a long healthspan." Varying the term makes for a more fluent translation, but the message can be stated with one translation: *support* R&D so that new technology can *support* your well-being.

The German word *Technik* has a range of meanings including both "techniques" performed and "technologies" implemented. In contemporary English, the key difference is between the *internalization* of knowledge as technique and the *externalization* of knowledge as technology. Philosopher Bernard Stiegler's *Technics and Time* differentiates between several meanings of *Technik*'s French cognate, *la technique*, through inflections of the word, but Stiegler's translators Richard Beardsworth and George Collins use the catch-all neologism "technics" to express the fundamental ambiguity of the term (Stiegler 1998, 280).

Like Stiegler's translators, I always rendered *Technologie* as "technology," but unlike them I was unwilling to resort to a neologism for *Technik* because Sorgner sometimes means techniques, sometimes technologies, and sometimes both, and I think these differences give a sense of the transhumanist's role in developing techniques that endow technologies with aspirational value. For instance, in a programmatic sentence from chapter 3, Sorgner writes: "Transhumanism embraces the use of technologies to increase the likelihood that posthumans may emerge" (*Transhumanismus bejaht den Gebrauch von Techniken, um die Wahrscheinlichkeit der Entstehung des Posthumanen zu erhöhen*). One could argue that Sorgner selected *Techniken*, not *Technologien*, because he is referring to the implementation of technologies, rather than their mere capabilities for potential use. However, what we normally call "technology" is at stake: these *Techniken* are as of yet external tools eventually to be integrated into the body.

In the chapter, techniques of self-improvement, like mindfulness training, are grouped alongside the technologies of self-enhancement that could result in humans' becoming capable of seeing UV light. "Techniques and technologies (*Techniken*) can both be means to change a human genome" (*Beide Techniken können Mittel sein, um ein menschliches Genom zu verändern*). The fact that human efforts to shape oneself through self-discipline are more familiar to most people than the tools that would permanently alter sensory experience make self-improvement "techniques" fully compatible with bioenhancement "technologies" in this context.

Stiegler's work presents the philosophical stakes of this loose concept's ambiguity. Consider sentences such as "Technics is the object of a history of techniques, beyond techniques. . . . Technics is not a fact, but a result" (Stiegler 1998, 30). We can only speak of technics by articulating the historical factors that produce a "technical system," which Stiegler defines as "a point of equilibrium concretized by a particular technology" (Stiegler 1998, 31). Particular technologies—like the steam engine, internal combustion engine, or the Bessemer smelting furnace—transform societies. These transformations have less to with inventive genius and more to do with response to a system of technological and economic possibilities, an analysis that Stiegler borrows from historian Bertrand Gille (Stiegler 1998, 35).

Anthropologie is tricky to translate because its meaning overlaps only partly with its English-language cognate. It is a philosophical term in German, whereas its English cognate designates only an academic field. *Anthropologie* has meant something like *theory of human existence* in German philosophy at least since Immanuel Kant lived and wrote. About Kant's idea of human dignity, Sorgner writes, "This term is an ontological one, since it implies a certain anthropology." German-language *Anthropologie* is also a discipline, one that answers questions about humanity's "position" (*Stellung*) in the universe, what distinguishes humans from other animals, and the extent to which biology can explain rationality. Its canonical founding fathers came from a variety of educational backgrounds: Helmut Plessner (1892–1985) from biology, Max Scheler (1874–1928) from theology, and Arnold Gehlen (1904–1976) from philosophy. Anthropology is a highly interdisciplinary field in the United States as well, but its bent is more distinctly empirical, at least since German immigrant Franz Boas (1858–1942) popularized the four fields approach (archaeology, linguistics, physical anthropology, and cultural anthropology). In German, *Anthropologie* still retains a speculative tenor. I often translate it as "theory of the human" to convey that context, especially when the word is preceded by an indefinite article.

The ordinary German word *Möglichkeit* can be translated unproblematically as "possibility," and that is how I usually translated it. The word often occurs in contexts about the "possibility" of human enhancements, where its meaning primarily suggests choice. But sometimes, as toward the end of the book in the discussion of Kevin Warwick's work, the idea is that there is, at least semantically, a latent potential in the technology itself, as opposed to possibilities available to a person: "Kevin Warwick's works clearly show the potential (*Möglichkeiten*) of the latest technologies." Earlier in the same chapter, Sorgner writes about three-parent fertilization and calls this a *Möglichkeit* because it is legally possible in Great Britain: "This technology (*Möglichkeit*) seems to offer an appealing option" to parents who might want that for disease-prevention reasons, in order to include two lesbian parents and a sperm donor, or for other reasons. Nevertheless, here too the syntax in English suggests that Sorgner is drawing our attention to a potential within the technology itself, and he thus deemphasizes the human agency that implements it. This was a particularly symptomatic ambiguity within a book about the affirmation of liberal choices regarding technology.

These subtle rhetorical relocations of agency from human to tool (akin to the transferred epithets of Homer, e.g., "Zeus's angry lightning bolt") are themselves not problems that can be trusted to machine translation. Nonetheless, as I describe below, the DeepL translation software had surprisingly good suggestions on the level of syntax, many of which I incorporated when translating this book. However, the software's strengths rarely extended to semantic distinctions, such as differentiating "technique" and "technology" in uses of the word *Technik*, and telling *Anthropologie*, the philosophical school of thought, apart from *Anthropologie*, the discipline associated with ethnography.

ENHANCED TRANSLATION

Like all writing, translation has always been a tool-assisted *technology* and is increasingly a computer-assisted one. Translation cannot be reduced to technology, however. As demonstrated in the examples of ambiguity above, it is also a labor-intensive *technique* based on extremely careful reading, a craft that improves with training and practice, and the work is still primarily performed by humans. About halfway through my work on the first draft of this translation project, I began to explore the functions of the Cologne-based DeepL Translator, a web-based machine-translation system. Appropriately enough, the book's author thoughtfully counseled me to try out this cutting-edge translation assistance tool to translate a book whose message was to embrace the expanding human-tool interface.[10]

For all of its strengths, tools like DeepL cannot replace a translator's care with such complex texts. Even the more personalized translation memory programs, like Trados, generally help translators only with the least rhetorically complex aspects of technical, legal, and medical texts. Software fails precisely in differentiating the tonal and collocational hallmarks of academic texts from those of other genres like literature, advertisements, or court documents. Machine translations of philosophy texts require time-intensive revision. Yet the embrace of translation software strikes me as transhumanist in the best sense: the labors of reading, understanding, and translation never occur solely in brains, and there is no pure reader to wall off from translation software's "artificial neural networks"—which perform many complex tasks besides translation, including detecting objects on a camera feed and diagnosing diseases. Artificial intelligence is transforming the style and economy of translation now that "deep learning" software is increasingly capable of learning new syntax and idioms. By recognizing patterns in one language, applications like DeepL can recognize word combinations even if the words appear separated within the sentence. Because the software draws on a large corpus of published texts in English, it can then rearrange the words in the proper order in the target language.

Our belief in the autonomy of individual minds (authors and translators) is difficult to relinquish, no matter how much we trust the insights of Freud or Kahneman (Freud 2003 [1920]; Kahneman 2011). Here is a heuristic analogy for understanding the role of the machine in my work on this book. *The Narrative of Arthur Gordon Pym of Nantucket* is written in the voice of the fictional character Pym. In the preface, Pym acknowledges the novella's actual author, Edgar Allen Poe, as the highly involved editor who wrote the beginning of the novella for Pym. Pym reports that he allowed Poe to "draw up, in his own words, a narrative of the earlier portion of my adventures . . . *under the garb of fiction*" (Poe 2008 [1838], 3). However, the author and the fictional narrator occupy different positions in relation to the novella: Poe has a reputation as an author of fiction, whereas Pym calls the very same work an authentic history. At the end of the introduction, Pym claims that the reader will be able to differentiate Poe's fiction from his own history simply because "the difference in point of style will be readily perceived." Though the reader is unlikely to notice that difference, the passage of time while reading a novel facilitates the suspension of disbelief: one forgets that one is reading a *fictional* story and becomes engaged with it, simply as a story. The same occurs with translations: when the reading experience is engrossing, we easily forget that we are reading a translation—we simply read.

Unlike this introduction, Poe introduces his novella in the voice of a fictional character and thus produces an irresolvable irony: a fictional character claims that he wrote the historically accurate part of the work and that an actual historical person wrote the fictional part. The introduction reads as a literary effect, a verisimilitude with no claim to historical veracity. To accommodate the transhumanist view here, I reserve the opposite judgment for machine-assisted translators. Although I translated as a cyborg, my human judgment ordered the entire process, and I thus do not expect readers to notice any "difference in point of style" in the passages where I considered machine input.

Computer languages can express a great deal, but computers have yet to demonstrate the capacity for the human experience of meaningfulness; for the computer, language functions only as a system of differences between words. The putatively "unsupervised" autonomy of machine learning cannot yet simulate a brain's openness to the manifold of experience. We machine-assisted translators are still performing the labor of translation ourselves, using foreign language skills acquired by *living* in languages, even as external technologies enhance our technique. Like the liberal transhumanist who wishes no one harm in his pursuit of an enhanced life, we cyborg translators hope that you will not judge our hybrid creations defective or inauthentic writing. The responsibility for any mistranslations is therefore mine alone.

—SPENCER HAWKINS, BERLIN, 2020

Introductory Remarks

"All transhumanists strive for immortality." "Transhumanism is a quasi-religious intellectual movement." Habermas goes as far as calling transhumanism "the worldview of a cult" (Habermas 2014, 36). These are among the prejudices circulating about transhumanism. Many are inaccurate. Transhumanism is not a cult. It is not even a religion. Transhumanism does not endorse prayer or ritual. It does not claim any demonstrable dogmas or religious symbols.

Transhumanists simply share one basic view, which they continually adapt to the latest state of philosophical insight, scientific research, and technological capacity. The view is that the use of technology has generally served human interests. From this point of view, they extrapolate from past effects of technology in order to envision a future where the appropriate use of technology might continue to transcend the limits placed on human existence, which would be in our interest since this would also support the probability of living a good life. Transhumanists accept the premise that humans emerged from evolutionary processes and could die out if the ongoing adaptation to ever-changing environmental conditions is unsuccessful. Constant self-overcoming is thus in our principal interest. It follows from this premise that most transhumanists adopt a naturalistic, nondualist, or relational theory of the human, which means that they consider it implausible to view human beings as consisting of an immaterial soul grafted onto a material body. Their view of humanity is a key reason why transhumanism has been called the most dangerous idea in the world.

2　　Considering that most transhumanists hold naturalistic views, serious-minded transhumanists consider it unthinkable that humans could achieve personal immortality—either in the sense of not having to die or of not being able to die. The basis for this insight can be clarified with numerous examples. We may succeed in flying to another solar system before our solar system dies in five billion years, but it is unlikely that a human being could survive when the universe collapses, for example, by the expansion process coming to a standstill or by a contraction following its current expansion where its total mass unites in one point of great density. A naturalistic view of human existence does not allow us to seriously consider personal immortality under such conditions.

Nevertheless, it must be admitted that numerous transhumanists do indeed talk about immortality. In such cases, immortality must be regarded as a utopia, which in turn must be taken as a rhetorical device. Most utopias have served this function in the history of philosophy. They are used to draw attention to a certain phenomenon, not because the philosophers presume that the utopias can ever be realized. The rhetoric of immortality here draws attention to the value of extending the lifespan. I prefer to speak of extending the healthspan since most people are not interested in just living longer, but in staying healthy longer, which could be realized by means of biotechnology, cryonics (cryogenic freezing of organisms in the hope that life can be continued in the future), or by mind uploading. Many recent events in human history make it clear that the healthspan is not a fixed quantity but a dynamic one.[1]

Although I have spoken thus far of a naturalistic image of humanity widespread among transhumanists, this statement must be qualified here at the outset. The transhumanists' naturalism is not necessarily a naïve naturalism that would entail the possibility of knowing truth as correspondence to reality. When naturalism is thought through seriously, it becomes necessary to affirm a variant of perspectivism as the most plausible methodology. This is an interpretive naturalism where the naturalistic worldview represents the most plausible interpretation of the world. It entails a rejection of realist ontologies and is tied to a methodological weakening. Even this methodological viewpoint, which stands in the tradition of Sextus Empiricus, Nietzsche, and Vattimo, is a reason to call this variant a weak transhumanism. To doubt the possibility of achieving truth in correspondence to reality does not mean that pragmatic concepts of truth can no longer be accepted. I have made this argument in detail elsewhere (Sorgner 2007; 2010b). Here I want only to argue briefly that many statements must be qualified. Especially in the examples discussed above, it is clear that widely held prejudices must be revised in a thoroughgoing engagement with transhumanism.

This is the central task of the present set of reflections on the most dangerous idea in the world. The organization is as follows: in the first chapter, "Is Transhumanism the Most Dangerous Idea in the World?" I introduce basic considerations about the history, goals, and cultural position of transhumanism, while asking whether transhumanism is indeed the most dangerous idea in the world. As discussed in the opening, transhumanists are interested in human self-overcoming and self-enhancement. Transhumanist perspectives thus play a central role in the scientific debates around enhancement technologies, issues that are introduced in their complexity in chapter 2, "A Roadmap of Enhancement Debates." These debates emerged when it became clear that medical technologies could do more than just fulfill therapeutic demands but could also help enhance relatively healthy lives. The chapter will evaluate the extent to which a clear conceptual distinction can be made between enhancement and therapy, and there are reasons to consider the distinction arbitrary. However, the importance of transhumanism extends beyond debates in applied ethics. Its cultural-historical meaning is best traced against the background of the history of the Enlightenment, which is why chapter 3, "Pedigrees of Metahumanism, Posthumanism, and Transhumanism" is dedicated to this spectrum of topics, as well as to the relationship of transhumanism to other contemporary philosophies of technology. One thinker whose work aims crucially at surpassing modernity and humanism is Friedrich Nietzsche. His significance for the twentieth and twenty-first centuries can hardly be overestimated. I show in chapter 4, "Nietzsche and Transhumanism," that Nietzsche's thought is structurally analogous to that of transhumanism, and I demonstrate that transhumanists can learn much from a critical response to Nietzsche's complex thinking. However, its significance for transhumanism goes beyond a complementary mentality. A

4

direct influence on transhumanism has taken place, as transhumanists like Max More, Zoltan Istvan, and me, too, were directly influenced by Nietzsche's reflections. Max More is one of the pioneers of futuristic thinking. Zoltan Istvan is a writer and politician, and since October 2014 he was even a US presidential candidate for the US-American Transhumanist Party. My own concern is a philosophical one.[2] In the concluding chapter, "Twelve Pillars of Transhumanist Discourse," I will clarify many of my basic philosophical views on a weak Nietzschean transhumanism and point out how they fit within the context of transhumanist discourses. These considerations make it clear that many widespread platitudes about transhumanism should be revisited. In any case, I hope that a scientific examination of transhumanism will also provide a fitting basis for public discussion of the relevant issues today.

It is not a realistic alternative to close your eyes and demand that technological advances come to a halt. If research cannot happen within nation-states, researchers will go to sea to pursue their scientific curiosity. This has a name: "seasteading." The first concern of this essay is therefore to foster new discussions on the norms, values, and anthropological implications of the latest technologies.

CHAPTER 1

Is Transhumanism the Most Dangerous Idea in the World?

In 2009, Nick Bostrom wrote an article entitled, "Why I Want to Be a Posthuman When I Grow Up."[1] Bostrom, cofounder of the World Transhumanist Association and philosophy professor at Oxford University, directs the Future of Humanity Institute. In that article, he argues that many people would be better off becoming posthuman with the help of new technologies. One key premise of transhumanism is the desirability of becoming posthuman.

The transhuman represents an intermediate form between the human and the posthuman. The transhuman possesses its own traits, which exceed human capacities, and transhumans therefore come into being as part of the posthuman's developmental process. Yet transhumanists themselves are still debating the exact meanings of transhuman and posthuman. Two positions prevail among transhumanists (Sorgner 2009):

Fereidoun M. Esfandiary, also known as FM-2030, takes it as a given that the transhuman is still a member of the human species, despite possessing traits so far beyond normal human ones that they constitute a bridge to the posthuman. The posthuman, for its part, would be a being that does not belong to the human species any more, and instead represents a further evolutionary step for humanity. FM-2030's line of thought shows structural analogies with Nietzsche's. Unlike FM-2030, Bostrom proposes the variant claim that posthumans would still belong to the human species while possessing traits that exceed those of humans living today, although he too argues that these special capacities need

6

not imply a new form of moral consideration, but that moral equality between humans and posthumans could emerge. The question of whether a moral equality would emerge is central for any evaluation of transhumanism. The novel *Brave New World* and the film *Gattaca* depict dystopian visions of the political problems that could accompany biotechnological developments.

The challenges for social justice exemplified in these stories are one reason Francis Fukuyama considers transhumanism the most dangerous idea in the world (Fukuyama 2004). Should we heed the warnings of the bioconservative political scientist and student of Alan Bloom, or might transhumanism represent the decisive intellectual orientation of the future, one that will make this a world worth living in, as transhumanists claim? These radically different views represent the extreme ends of the spectrum, especially within English-language discussions. In German-speaking countries, by contrast, transhumanism has not received much attention either in academic or in public discussions. This may result from the fact that a bioconservative stance prevails with regard to human biotechnology research and to the application of innovative research findings in that area. Transhumanism represents the bioliberal opposition to this verdict. Beyond that, it represents the most radical embrace of scientific, medical, and technological developments.

The most prominent statement by a German intellectual on transhumanism occurs in Jürgen Habermas's text *The Future of Human Nature*, which characterizes the transhumanist movement as follows:

> A handful of freaked-out intellectuals is busy reading the tea leaves of a naturalistic version of posthumanism, only to give, at what they suppose to be a time-wall, one more spin—"hyper-modernity" against "hypermorality"—to the all-too-familiar motives of a very German ideology. Fortunately, the elitist dismissals of the "illusion of egalitarianism" and the discourse of justice still lack the power for large-scale infection. Self-styled Nietzscheans, indulging the fantasies of the "battle between large-scale and small-scale man-breeders" as "the fundamental conflict of all future," and encouraging the "main cultural factions" to "exercise the power of selection which they have

actually gained," have, so far, only succeeded in staging a media spectacle. (2003, 22)

The naturalistic version of posthumanism to which Habermas refers is transhumanism, and he lumps it together with the topic of Peter Sloterdijk's widely received 1999 talk entitled, "Rules for the Human Zoo: A Response to the Letter on Humanism." Sloterdijk's talk gave decisive stimulus to the debate on human biotechnologies at the end of the 1990s. The phrases in quotation marks in Habermas's text come from Sloterdijk's talk on the "human zoo," and, as those references suggest, Habermas's text is an implicit answer to Sloterdijk's talk.

The above quoted passage by Habermas misses the mark in several respects:

1. The phenomena he writes about are not posthumanist and are not even "a naturalistic version of posthumanism"; rather, they are transhumanist.
2. Sloterdijk is not a transhumanist.
3. There is reason to doubt that transhumanism lacks "the power for large-scale infection."

Each of these objections merit further elaboration.

Objection 1: I will present a more precise differentiation of trans- and posthumanism in the chapter entitled "Pedigrees of Metahumanism, Posthumanism, and Transhumanism." This range of concepts is discussed in detail within the introductory guide *Posthumanism and Transhumanism: An Introduction* (Ranisch and Sorgner 2014). One hallmark of posthumanism is its indissoluble bond to the tradition of Continental philosophy, whereas transhumanism has prospered in the naturalistic and utilitarian thought of the English-speaking world.

Objection 2: Sloterdijk's talk, "Rules for the Human Zoo," solely argues for the necessity of reflecting on the ethical questions raised by human biotechnologies. There are only two reasons why Habermas would find his talk disconcerting: the talk draws on Plato, Nietzsche, and Heidegger, who are suspected within the German-speaking world of advocating a totalitarian and fascistic line of thought (for the most part unjustifiably, in my view), and the talk takes up a vocabulary (human zoo, breeding) that hardly softens that impression. Sloterdijk has not uttered any affirmative moral judgments on the enhancement of humanity. On December 6, 2005, Sloterdijk gave a presentation at the University of Tübingen called *Optimierung des Menschen?* (Optimization of the human?),[2] where he made it explicit that he, just like Habermas, views the application of human biotechnologies for therapeutic purposes as morally suitable but sees it as morally problematic to use them for enhancement purposes. That view means that he cannot be a transhumanist. By definition, transhumanists embrace the use of enhancement technologies.

Objection 3: Perhaps we cannot speak of transhumanism's "power for large-scale infection" in German-speaking countries. There are however good reasons to believe that this lack of interest does not hold in an international context. Several facts speak for the opposing view: the large volume of academic publications that engage with transhumanist positions, especially in the fields of bioethics and medical ethics, the fact that leading transhumanists hold positions at top universities in the English-speaking world (such as Oxford University), and the intensive, multifaceted engagement with questions of transhumanism in the artistic and cultural domain. Popular engagement with transhumanism proliferates in the form of well-known films (*Gattaca, Transcendence*), novels (Michel Houellebecq's *The Elementary Particles* and *The Possibility of an Island*, Dan Brown's *Inferno*, Zoltan Istvan's *The Transhumanist Wager*), visual art (Patricia Piccini's *Still Life with Stem Cells* or *Alba*, the fluorescent rabbit by Eduardo Kac), and science fiction literature in general.

Habermas also makes the provocative claim that transhumanists are a bunch of "freaked-out intellectuals" with Nietzschean fantasies. This portrayal is controversial among transhumanists, probably because they are aware of Friedrich Nietzsche's murky reputation, and they don't want association with Nietzsche to bring them under suspicion of supporting morally reprehensible views.[3]

After this brief background in transhumanism's neglect in the German-speaking world, the next question is whether increased engagement with this intellectual movement could exert a decisive influence on German-language debates on bioethics and medical ethics or on the philosophy of technology. Detailed engagement with this question requires a more nuanced account of the movement. I will thus introduce some further facets of transhumanist thought below.

FACETS OF TRANSHUMANISM

The use of the concept "transhumanism" was first coined by Julian Huxley in the 1951 article, "Knowledge, Morality, and Destiny,"[4] and then developed in his 1957 book *New Bottles for New Wine*. The author was the brother of Aldous Huxley, the author of *Brave New World* (1932) and the uncle of "Darwin's Bulldog" Thomas Henry Huxley. The novel *Brave New World* represents a direct response to notions propagated by Julian Huxley, which he already advocated before he developed the concept of transhumanism.

For the contemporary understanding of the concept, the most decisive work is the book, *Are You a Transhuman?* published in 1989 by FM-2030, who used to be called F. M. Esfandiary but renamed himself FM-2030 to draw attention to the arbitrariness of name assignment and to announce his wish to celebrate his one hundredth birthday in the year 2030. He already died in 2000 from complications resulting from a pancreatic tumor. Max More's article, "Transhumanism: A Futurist Philosophy," originally published in 1990, also exerted a formative influence on the contemporary understanding of the concept. Max More was born Max T. O'Connor and changed his last name to More as a reminder of the importance of self-overcoming—humans always want to achieve more and to overcome their limitations constantly, in his view, and thus he chose the evocative name, "more." FM-2030 and Max More have more in common than having established the contemporary concept of transhumanism and having changed their own names to match their worldviews. They also both had ties to the transhumanist artist Natasha Vita-More. She and Max More are still married.

A further important step in the history of transhumanism was the founding of the World Transhumanist Association by Nick Bostrom and David Pearce in 1998, along with the political debate among transhumanists, which occurred in reaction to James Hughes's book, *Citizen Cyborg*. Hughes takes on a social-democratic position in opposition to More's extropic variety of transhumanism (extropy—the opposite of entropy—is used in a metaphoric sense), which tends much more toward libertarian stances. This discussion continues today when bioethicists take up the so-called Gattaca argument. Bioethicists reflect on whether the further development of enhancement technologies would lead to a segmentation of society into the gene rich and the gene poor, or the posthumans and the humans. They also think about whether belonging to one group might come with different rights than belonging to the other.

There are two ways to explain why social division would not necessarily stand as a threat. Consider the analogy between vaccinations and other enhancement technologies; vaccinations are after all a kind of biotechnological enhancement. There had been a legally sanctioned vaccination requirement in Germany until the 1980s, but it is no longer in effect. Vaccinations, which are reliable, helpful, and safe for many people, are covered by health insurance if desired. Health insurance even pays for special vaccinations in some cases. There are however many unusual vaccinations that individuals must pay for out of pocket. Analogously, the safe and helpful enhancement technologies could be universally available in the same way, in order to ensure that a segmentation of society does not occur.

Here is the second reason why the Gattaca danger need not loom. Thirty years ago, only business leaders could afford cell phones. Today almost every member of Western society could potentially own and use such a phone. This example illustrates how successful technologies that are helpful, reliable, useful, and safe are quickly in demand by so many people that the technologies soon become more affordable and within reach of many people. We could expect an analogous development with other enhancement technologies, which would prevent a segmentation of society into groups of different value. If the consequences described in this scenario did occur, however, the next concern would be that citizens who do not wish to use these technologies would still be forced to use them. This judgment may be correct, but it need not be a criticism as the following example emphasizes. Thirty years ago, students were still permitted to submit hand-written papers. This is no longer acceptable today since it is generally expected that they are written on computers and formatted in a specific way. Students are thus compelled to use computers. Is this fact morally problematic? Not necessarily. For many people, computers are affordable, reliable, safe, and useful. The advantages clearly outweigh the disadvantages. Students hardly ever find themselves wishing to write a paper without a computer. If a lot of enhancement technologies are inclined to develop in just this way, then it is likely that the consequences would be analogous to those in the aforementioned case. The reasons sketched here make it clear that the splintering of society into a Gattaca-like structure need not necessarily accompany the further development of biotechnologies. With this reassurance, we can enjoy the rewards of engaging more deeply with transhumanism. Such engagement need not be morally dubious work.

Even before Julian Huxley constructed the notion of transhumanism, there were already thinkers who worked on topics that would later be decisively integrated into transhumanism. Especially noteworthy are the utilitarians, some Enlightenment thinkers, Darwin, and Nietzsche—although this ancestry is controversial. The goals and basic views of transhumanism rely on a naturalistic notion of humanity, which differs by degree, not by type, from other organisms. Transhumanist views often, but not necessarily, imply a utilitarian ethical position and liberal-democratic social order; there are however libertarian and social-democratic branches of the movement.

The decisive feature of transhumanism is its advocacy for new technologies to increase the probability that transhumans or posthumans will emerge, so that evolution no longer depends on natural selection but can let human selection set its course. This development deeply calls into question whether humanity has reached its maturity. As small children depend on their surroundings and only begin to become masters of their own lives as teenagers, so human evolution has depended on natural selection. Now humanity is slowly coming to its evolutionary teens, where we are ever readier to take control over our own evolutionary process. The analogy has limited scope; it only represents an attempt to approximate transhumanist philosophy.

The strong embrace of radical enhancement technologies is the defining feature of transhumanism and represents the movement's most innovative aspect. Transhumanists especially support the enhancement of emotional, physical, and intellectual abilities, along with the

12

extension of the healthspan during which the transhuman or posthuman can emerge (Bostrom 2009, 113–16). The transhuman or posthuman is not only about a new description of humanity, not just a new anthropological position, but rather it is about new developments in humanity. This movement is characterized by an optimistic view of the future and of the further development of humanity.

Transhumanists offer one main premise for arguing that surpassing humanity is desirable: individuals will regard their own quality of life as higher when their emotional, physical, and intellectual abilities are enhanced and their healthspans are extended. Whoever commands greater capacities and remains alive longer in a healthy condition generally leads a more comfortable life than those for whom this is not the case. A lot of transhumanists support this claim with evidence from psychological research (Bostrom 2009, 116). Do we indeed become happier when we develop and apply new enhancement technologies?

The traits and capacities that are especially relevant for fostering a good life are emotional, psychological, and intellectual capacities, along with a long healthspan. Lately, the leading enhancement technology researchers and transhumanists are considering the possibility of pharmacologically improving morality.

Which technologies should be prioritized from the transhumanist point of view? Based on present concerns, the following four areas of enhancement technology are discussed below: 1. genetic enhancement; 2. pharmacological enhancement, that is, medications, drugs for doping or recreation—in short, neuroenhancement; 3. cyborg enhancement by establishing interconnected human-machines, commonly known as cyborgs (cybernetic organisms), that is, complex systems that combine living organisms with digital or nondigital machines, which would allow humans to link up with digital as well as mechanical machines; 4. morphological enhancement, that is, beautification surgery.

THE EVALUATION OF TRANSHUMANISM

Is transhumanism the most dangerous idea in the world? Transhumanists embrace the liberal-democratic order as foundational and thus attach great importance to the norms of freedom and equality; Max More focuses on freedom, whereas free and equal rights are the focus for James Hughes. I for one consider both of these standards *the* key achievements of the Enlightenment worth defending. Movements directed against these basic standards are undoubtedly much more dangerous than transhumanism. This is not to say that no dangers can be found in transhumanist ideas.

A common reaction to transhumanist thinking is spontaneous disgust. Just thinking about self-modification makes many people go "Yuck!" The American bioethicist Leon Kass even believes that there is wisdom in such a reaction, the wisdom of revulsion contained in the yuck factor (Kass 1997). However, an emotional, negative reaction does not seem to me to arrive at an adequate or plausible assessment of transhumanism. Telling another narrative of the possibilities arising from human biotechnologies can provoke other emotional reactions to transhumanist proposals, such as an affirmative "Yeah!" Modifying a human being need not be disgusting. 250 Two hundred and fifty years ago, the enhancement of humans by vaccination had not yet been developed. "Yeah!" Two hundred years ago, there were no anesthetics yet. I find it terrific that we can now use anesthetics. "Yeah!" One hundred and fifty years ago, there were no antibiotics yet. I am very grateful that they are now available. "Yeah!"

In fact, progress in biotechnology is linked to many remarkable achievements. Of course, every technical innovation also comes with new dangers. Transhumanists are well aware of these dangers, and they are intensively grappling with them. Does this mean that there are no moral problems associated with transhumanism? This is hardly the case, as some transhumanists even speak of enhancement as a moral duty, and they even apply this obligation to the processes of genetic enhancement. For example, they argue for the moral duty to select and implant the fertilized egg after in vitro fertilization and preimplantation genetic diagnosis that has the highest chance of living a good life. Not only transhumanists uphold this assessment at present. The bioliberal director of the Uehiro Center of the University of Oxford, Julian Savulescu, propounds this thesis, which I consider to be morally problematic, since it is inextricable from certain paternalistic tendencies

(2001, 413–26; Sorgner 2014a). It certainly resembles transhumanism though.

Within this section, I was primarily aiming to explain transhumanism and to put its tenets up for discussion. To disregard the role of transhumanist considerations from the bio-, medical-, and technical-ethical discourses would amount to a lack of international outlook. Engaging with thoughts does not imply agreement with the thoughts under discussion. Transhumanism involves considerations that challenge widespread prejudice in an intellectually stimulating way and encourages us to think further.

Do I want to be a posthuman when I get older, like Bostrom does? That is not how I would formulate the objectives for my life planning, but I do believe that constant self-overcoming is central to promoting my own quality of life. I also consider scientific research, especially in biotechnology, extremely important and advocate for greater sponsorship of those research fields. I consider the availability of anesthetics, vaccinations, and antibiotics important achievements. I hope that future achievements will also address important challenges. This stance can be parsed as a weak form of transhumanism.[5]

CHAPTER 2

A Roadmap of Enhancement Debates

Every new technology leads to hitherto unknown challenges; this was the case for the development of steam ships and aircrafts. Many experts herald the dawning twenty-first century as the "century of biology" (Sorgner 2006a, 11). Innovations in biotechnology raise new ethical questions—perhaps more so than any other recent developments. Decisive developments in biotechnologies not only have helped cure diseases but are already blowing away the previous limitations on what it means to be human (Eßmann, Bittner, and Baltes 2011; Hauskeller 2009).

Humans are in a position to transform themselves, can actively intervene in evolution, and can thus become, not unlike a modern Prometheus, "creators of humanity" (Sorgner 2011, 19–21). Here some basic questions emerge: What are the ethical limits for such interventions in medicine? Can and should we strive to improve on human emotions and morals?

In the first part of this chapter, I will present some of the capacities whose enhancement is currently under intense debate (Sorgner 2006a, 11). In the second part of the chapter, I will introduce a selection of the techniques that promote human skills. This should clarify the philosophical context around the question of the enhancement of emotion and morality, which I want to grapple with in the final third part. In this context, I am interested not in reaching a final normative assessment of the various techniques but rather in situating the techniques historically and giving a rough overview of ethically relevant issues for a dynamically developing field (see, inter alia, Savulescu, Meulen, and Kahane 2011).

HUMAN CAPACITIES

What ought to be "improved" or "enhanced?" Which capacities are mentioned with particular frequency in scholarly discussions about enhancement? It makes a significant difference for moral assessment which ones are under primary consideration for modification (eye color, size, sex, erotic orientation, intelligence, lifespan). Within this chapter, I focus on the areas that are at the heart of current bioethics debates. It should be noted that emotions, physiological or intellectual capacities, and healthspan are the most intensely discussed areas in this context (Bostrom 2009, 108). In the recent past, leading enhancement researchers have also considered the possibility of the pharmacological enhancement of morality. In the following overview, I outline selected considerations that are intended to give a first impression of how we might discuss the enhancement of various types of human capacity.

Enhancement of Emotional Capacities
I will discuss the topic of emotional enhancement in more detail in later sections (Krämer 2009; Sousa 2009; Stephan 2004; Stephan and Walter 2004). The following is meant to illustrate intuitively that the developmental acquisition of new emotional capacities is familiar to everyone, which illustrates that an analogous process could also be possible with respect to other emotions; this is to say, our emotional disposition could be quite a variable part of who we are (Bittner 2011). A toddler has no template for romantic love; this feeling first comes into fruition during the teenage years. That is a clear example of our ability to acquire new emotions. With the help of various enhancement technologies, it could be possible to access new feelings as adults and to amplify the presently existing ones so that human life on the emotional level is extended to new horizons, which could make human life even richer, more varied, and more exhilarating (Savulescu and Sandberg 2008).

Enhancement of Physiological and Intellectual Capacities
There are many fascinating examples worth mentioning here. Noteworthy in particular are cases of "savantism." So-called savants can have exceptional memories or exceptional talents in linguistics, music, or other arts.

Take the case of a person with savantism of memory. Kim Peek, model for the protagonist of the film *Rain Man*, was able to read two pages of a book at the same time, taking in one page with each of his two eyes; he knew the contents of a total of around twelve thousand books by heart (Siefer 2009, 37). His abilities were, of course, not limited to this feat. However, this particular example alone shows what the human brain can accomplish. If a brain is able to do that, it cannot be ruled out that, with the aid of enhancement technologies, measures can be taken that enable other brains as well to yield other similar accomplishments.[1]

We could ask whether it is desirable to possess such capacities. Bio-technology advocates tend to answer that question as follows: as small children, many of us may have led safe and happy lives. As adults, our lives are far less carefree and peaceful, and yet we possess capacities that we lacked as children. Many adults value their capacities so highly that they would not wish to be children again if such a transformation were possible. John Stuart Mill proposes a similar consideration in the second chapter of *Utilitarianism*: "Better to be Socrates dissatisfied than a fool satisfied" (Mill 2006). This consideration seems to suggest that possessing intellectual and physiological capacities has an intrinsic value. Psychological studies suggest that this is true for many individuals. Universal approval cannot, of course, be presumed (Bostrom 2009, 115, 116–18.).

Enhancement of Healthspan

Numerous liberal bioethicists do not consider extending the lifespan a worthy goal (Hilt, Jordan, and Frewer 2010) because extending the lifespan does not necessarily entail a sustained quality of life (Bostrom 2009, 113–16). A long breakdown at the end of life would not be considered desirable by very many people. Improving the healthspan implies that a person does not just live longer but stays healthy longer (Hilt, Jordan, and Frewer 2010). The view of aging that corresponds with this notion is one where the various phases of life extend proportionally to the present sequence. The time between being a newborn and a teenager could last not just twelve but twenty-four years. According to this model, people could live in relatively good health until their one hundredth year, and then experience a relatively short period of decline. Would the extension of healthspan foster a high quality of human life?

This question cannot be answered with a simple, "yes." Bostrom has dealt with this question in an article, and he has proposed, in reference to psychological and other empirical studies, that the probability of having a good life increases with the technologically mediated extension of the healthspan (Bostrom 2009, 116).

Aubrey de Grey follows this line of thought and considers aging an illness that we have a responsibility to fight. His studies have crystallized the "deadly sins of aging," and he founded the Methuselah Foundation to fight them (de Grey and Mae 2007).[2]

Some means of extending healthspan and of increasing emotional, physiological, and intellectual capacities are already available, and these are practiced by many segments of the population already. Prominent examples include the use of stimulants, sedatives, and concentration-enhancing substances, as well as cosmetic surgeries (Devereaux 2009; Schöne-Seifert et al. 2009).

TECHNOLOGIES FOR ENHANCING EMOTIONS AND MORALS

Which technologies might bolster the aforementioned capacities? Genetic modification, consuming pharmaceutical products, and constructing human-machine hybrids, known as cyborgs (cybernetic organisms), are the technologies most often found in scientific accounts (Savulescu, Meulen, and Kahane 2011). In the next sections, I will analyze some issues surrounding those technologies. I will discuss the four central enhancement technologies briefly: genetic enhancement; pharmacological enhancement by taking chemical and biological substances (e.g., medications, "doping" supplements, or recreational drugs); cyborg-enhancements; and morphological enhancements (i.e., cosmetic surgeries), which represent the most economically successful of the enhancement technologies.

Genetic Enhancement

In the German-speaking world, questions of genetic enhancement came to the fore in the public consciousness with Peter Sloterdijk's "Rules for the Human Zoo," a controversial speech delivered in Elmau (Sloterdijk

2009). We can take Habermas's *The Future of Human Nature* as an indirect answer to that speech (Habermas 2003). Among German-speaking philosophers, Habermas has produced the internationally prominent statement on genetic enhancement or liberal eugenics. Habermas takes on a bioconservative position. In the international context, we can name thinkers like Kass, Sandel, and Fukuyama as important adherents of such a basic position. Their opponents include Harris, Savulescu, and numerous transhumanists (Savulescu and Bostrom 2009). Transhumanists differ from bioliberals like Harris in that they explicitly view the emergence of the posthuman as desirable.

In the late 1990s, the term "liberal eugenics" was the preferred heading for the topic of fostering hereditary dispositions. Liberal eugenics can be sharply distinguished from the state eugenics practiced in the Third Reich. In state eugenics, the state decides how and in what form hereditary dispositions should be enhanced, and no other form of eugenics are allowed. In liberal eugenics, by contrast, a person decides on those matters individually (autonomous liberal eugenics), or parents decide for children (heteronomous liberal eugenics). Since the word "eugenics" is still associated with the most morally problematic actions, advocates of tantamount measures have established the term "genetic enhancement," which still means nothing other than improving one's genetic dispositions. The use of the word "enhancement" has thoroughly permeated the discourse. Those ethicists who use the word "eugenics" tend to be bioconservatives, who do so consciously in order to call up certain associations. The concepts of liberal eugenics, or genetic enhancement, call up technologies that not only directly modify a human genetically, but more prominently they also deal with genetic enhancement by selection measures, like the selection and implantation of fertilized egg cells (intrauterine fertilization) and preimplantation diagnostics. Many bioliberals and transhumanists pronounce it a moral duty to arrive at genetic enhancement by selection. Work in synthetic biology shows great overlap with the possibility of genetic enhancement.[3]

Pharmacological Enhancement

The moral evaluation and engagement with questions around doping have long been discussed as a matter of sports ethics. Savulescu has commented extensively on the topic (Foddy and Savulescu 2009). A further aspect of this topic is discussed under the headings of neuroethics and neuroenhancement (Schöne-Seifert 2009). The latter includes moral considerations about managing the brain with psychoactive substances, including: (1) Donepezil, which enhances memory and thought capacity, for example in patients with Alzheimer's; (2) Ritalin, which enhances concentration abilities, for example in patients with ADHD; (3) Modafinil, which regulates circadian rhythms and alertness, for example in patients with narcolepsy (Schöne-Seifert et al. 2009, inter alia). And of course let us not forget the most successful pharmacological enhancement technology since its release in 1998: Viagra.

Cyborg Enhancement

In the recent history of cyborg enhancement, Kevin Warwick's work has been crucially important; he has done pathbreaking research in robotics and in linking the human nervous system with computer systems (Warwick 2012). Ray Kurzweil's thinking and research are also relevant (Kurzweil 2005). When cyborg technology comes up, so does the possibility of uploading the contents of human brains onto computers, and that very possibility raises fundamental questions about what it means to be human. We also have to consider whether personality and the rights that accompany it can be attributed to computers (Kurzweil 2005). This line of thinking invites questions about which traits define what it is to be human, and which ones should define it. In sports ethics, the cyborg is discussed especially because of the achievements of the runner Oscar Pistorius, whose legs were amputated below the knee and who runs with prosthetic legs, sometimes faster than the best sprinters. We will see such discussion of cyborg technology further in the future (Hildt and Engels 2009).[4]

Morphological Enhancement

The economic significance of morphological enhancement has risen sharply in recent years. The following procedures, considered cosmetic enhancements, have been the most significant: liposuction, breast augmentation, rhinoplasty, face lifts, and wrinkle removal. Beauty operations have developed to a high degree in South Korea. Nose and eyelid surgery to look more European are especially popular.

In Germany, liposuction, breast augmentation, and face lifts are especially prevalent. These procedures are more socially acceptable now than they were ten years ago. The change has caused body parts to be observed aesthetically that were formerly not submitted to scrutiny. For example, the form and shape of vaginal lips can now be modified aesthetically in order to match the image of a visual norm promulgated by pornography. Hymen reconstruction surgery is a technology that could be seen as even more problematic from a social-ethical perspective.

ENHANCEMENT OF EMOTIONS AND MORALS

One cannot overstate the significance of consuming medications and establishing human-machine hybrids as means of emotional modification. A decisive question in this context is what the goal of enhancing emotions is. Two domains emerge, which are intensively discussed in the literature: 1. enhancing emotions for the promotion of the good life, 2. enhancing emotions to promote morality. The second of these is by far the most controversial and questionable area.

Enhancing Emotions to Promote the Good Life
This is the area concerned with individual well-being. Examples include patients with depression taking antidepressants, and healthy patients taking mood-enhancing medications like Prozac (Schmidt-Felzmann 2009; DeGrazia 2004; Elliott and Chambers 2004). When we speak of promoting the good life, then what we are promoting only concerns individual well-being, that is, the personal judgment of one's own satisfaction. Interrogating the good life means asking which virtues, attitudes, abilities, and insights are necessarily connected with living well; this assumes that generalizable answers can be given to these questions. In ancient and medieval philosophy, this notion was treated under the heading of *eudaimonia* and stood at the center of most schools of thought. After the Enlightenment, a basic position has risen to dominance that sees it as implausible to arrive at one answer to the question

of the conditions of happiness. In contemporary philosophy, a few communitarian thinkers, like Martha Nussbaum and Michael Sandel, are engaged with precisely that—even if these thinkers, unlike their ancient predecessors, are only describing a strong but vague concept of the good. I consider their project dangerous because paternalistic and potentially totalitarian structures accompany it.

Enhancing Emotions to Promote Morality

The domain of moral enhancement dispenses with questions of individual well-being and takes on the possibility of promoting morality by enhancing emotions, which could occur both through human-machine hybrids, such as in deep brain stimulation, and through consuming medications (Schöne-Seifert et al. 2009). Morality includes concepts like the morally right or just (*[Ge-]Recht*) (Sorgner 2010b, 13). The question arises as to whether correct behavior can be achieved by moral enhancement technologies. Tom Douglas, Ingmar Persson, and Julian Savulescu believe that it can (Persson and Savulescu 2011; Douglas 2011). I personally am skeptical of the hypothetical possibility of promoting or modifying morality because I can hardly imagine how the norms of freedom and equality could be achieved by creating cyborg technologies or taking medications. This topic has become relevant in debates about various enhancement technologies. When the variety of possible enhancement technologies became clear, the challenge of abusing these technologies emerged. Shouldn't people be morally enhanced before they are confronted with the potential effects of enhancement technologies? The problem that emerges with the expanded power of enhancement technologies is that of global destruction. For example, a scientist could create a virus that is activated only after a long delay but then leads to a sudden death. The ethicists Ingmar Persson and Julian Savulescu have engaged especially with the matter of global catastrophe and the need for moral enhancements (Persson and Savulescu 2012). Saving the human species from extinction may require us to address the question of moral enhancements.

Technologies for Enhancing Emotions

Technologies repeatedly mentioned in this context include the following:

1. Deep brain stimulation is used to treat Parkinson's disease and major depression (Schlaepfer and Lieb 2005; Müller and Christen 2010).

2. Antidepressants like amitriptyline enhance individual mood and are used to treat depression, anxiety, and other mood disorders, as well as to alleviate chronic pain.

3. Oxytocin enhances trust of others and is used, for instance, with social phobias and autism, or when mothers feel detached from their children (Hurlemann et al. 2010).

These technologies serve not only therapeutic ends; their use can also assist healthy people to enhance certain states, qualities, and abilities. It is a completely open question whether the distinction between therapy and enhancement is tenable. Generally, the following tendency can be ascertained: opponents of enhancement primarily hold fast to the distinction, where they view therapies as morally sound, but they do not hold the same view of enhancement measures. Enhancement advocates by contrast emphasize the fluid, unclear passage from therapeutic to improving or "enhancing" measures and (often) conclude from there that the distinction should be abandoned. Furthermore, they emphasize that enhancement technologies are not necessarily morally reprehensible since the enhancement of abilities is not morally wrong, and employing means of doing so with children (*zu Erziehenden*) is structurally analogous with education itself (*Erziehungsmethoden*) (Chadwick 2009; Sorgner 2010a, 2015). Beyond that, many models emphasize the right to morphological freedom as a way to explain why there should be free access to these technologies; that is, the state should not forbid the use of these technologies. The following can be argued regarding some technologies and substances: since the side effects of Prozac and cigarettes ought to be compared, if it were found that Prozac had less severe side effects than cigarettes, then Prozac ought to be marketed and sold at least as freely as cigarettes.

Discussions also attempt to gauge the effects of these technologies on healthy people, but studies in this area show no unified results. Some studies attribute no efficacy to these drugs when administered to healthy individuals, whereas other studies report the opposite; other studies suggest that neuroenhancers enhance the abilities of individuals with low ability but have no decisive supplementary effect on individuals with high ability (Stix 2010, 50–51).

An interesting factor in these discussions is how widely dispersed these enhancement technologies already are in any particular society. German Employee Health Insurance reported the following about health in Germany in 2009: according to a survey, 5 percent of employed Germans between twenty and fifty take medications to enhance performance or mood (DAK Gesundheit 2009). Similar surveys were given at US universities, and these showed even higher rates of consumption of performance-enhancing medications, whereas the rates are lower at German universities (Stadt Zürich, n.d.). More recent studies show hardly any difference between the rates of performance-enhancing substances at German and American universities and show increased consumption rates overall in comparison with past years (Müller-Jung 2013). If this data is accurate, then it suggests a social justice problem for biotechnology research regarding the just distribution of resources. The question is whether students who medicate are not just as morally culpable as athletes who dope.

If the aforementioned technologies are indeed effective and reliable, then we might also ask whether it would make sense to fund these medications through public health insurance companies. If biotechnologies foster either individual well-being or moral constitution, then are there not structural analogies between these and technologies like vaccination? Would it not make sense to proceed here as with vaccinations? As noted earlier, vaccinations are voluntary in Germany, where they have not been mandatory since the 1980s. Many vaccinations are covered by health insurance, others are covered under certain conditions, while others yet must be paid for by individuals. Wouldn't a similar manner of proceeding make sense in the case of many other enhancement technologies?

A further challenge raised by enhancement technologies is that of maintaining individual identity. Am I still myself once I take certain antidepressants? After consuming certain amounts of alcohol, I am no longer "myself" or no longer as responsible for my behavior since I am no longer in full self-control or legally culpable. Section twenty of the German Criminal Code states that an individual "who at the time of the commission of the offense is incapable of appreciating the unlawfulness of their actions or of acting in accordance with any such appreciation due to a pathological mental disorder, a profound consciousness disorder, debility or any other serious mental abnormality, shall be deemed to act without guilt" (Bundesamt für Justiz, n.d.; see Chadwick 2009; Sorgner 2010a). Does individual identity and culpability vanish when we use certain emotional enhancement technologies, or does that occur only after we have reached a certain threshold of use?

A lack of impulse control is an example of a "pathological mental disorder." According to the German Criminal Code, people with such a disorder are not culpable. If such people agree to a pharmacological treatment for such a disorder, for example, pharmacological castration of sexual perpetrators, then this disorder is removed, and they become culpable again. Have they then "reunited with themselves?" Could such measures also be taken with people who have no disorders, as a way of amplifying their abilities, if that is how they see taking measures to prevent future loss of impulse control?

Could mothers who do not accept their children be given oxytocin to relieve their condition? Would mothers be more "themselves" through this medication, or would they just be more "well-adjusted," so that they fulfill their roles as mothers more appropriately? If socially undesirable behavior is seen as pathological, then oxytocin could be mixed into the public drinking water. I do not want to be misunderstood here. I do not think that such a behavior is morally justified or could ever be. I just want to show the central dimensions of the questions raised here.

The side effects of deep brain stimulation, an experimental treatment for depression, could hypothetically make it necessary to take measures for moral enhancement. Patients who undergo deep brain stimulation experience short-lived manic symptoms in the worst-case scenario (Ulla et al. 2006). These undesirable side effects could require treatment by moral enhancement technologies.

In the previous sections, I have primarily concentrated on the enhancement of emotions by pharmaceuticals or cyborg technologies. It cannot be overlooked, however, that this kind of enhancement could also be achieved through the selection or modification of genetic makeup. The following example illustrates a case for legalizing the possibility of enhancing emotions through genetic modification: Dean Hamer of the National Cancer Institute in Bethesda announced that he had discovered a gene responsible for homosexuality (Hamer et al. 1993). The bearer of such a gene supposedly has a high likelihood of being homosexual. There are of course people with the gene who are not homosexual, and people without it who are. The correlation between the presence of this gene and homosexuality is, however, considered high. Let us imagine that such a gene exists. This gene, when present, influences the emotional orientation decisively toward members of the same sex. Through selection after in vitro fertilization (IVF) and preimplantation genetic diagnosis (PGD) or genetic modification technologies, it is possible to influence the presence or absence of a gene—which could certainly influence a person's emotional orientation. It is an ethical question if and to what extent such influence is morally legitimate. If this description is appropriate, it could be that such technologies are being selected (after IVF and PGD) so that people may be born with or without a gene that affects sexual orientation, or that this gene could be added, activated, removed, or deactivated by means of genome editing. This knowledge could make it possible for parents to decide whether a child is likely to be born homosexual or heterosexual, and to change a child from one orientation to the other. All of this is to say that the possibility of modifying emotions could be relevant even though this area is hardly at the center of research, and the possibility of such modifications remains hypothetical.

One core branch of current research on emotional enhancement focuses on achieving moral enhancement by influencing serotonin release. A debate on this work emerged through the research of neuroscientist Molly Crockett, who wrote her dissertation on this at the University of Cambridge (Crockett et al. 2010). Her research showed that giving citalopram to healthy trial participants increased serotonin release. It then emerged that, with higher serotonin levels, volunteers were less inclined to harm others, from which she concluded that serotonin contributes to "prosocial behavior." She measured this tendency among trial participants with two experiments: the Ultimatum Game and a variant on the Trolley Problem, known as the Fat Man thought experiment.

The Ultimatum Game is often used in experiments conducted by economists. The experiment is structured as follows: two players interact in order to decide how to divide a quantity of money given to one of the players. Player A is allowed to decide how to divide the money; player B can either accept or reject the offer. If player B rejects the offer, neither player gets the money. If player B accepts it, then the money is divided as proposed. The game is played only once so that no reciprocal dependence of behavior can emerge. The results were as follows: the higher the serotonin levels of the participants, the more inclined they were to accept the offer, even if it was not very fair. Crockett concluded from this that the elevated tendency of participants not to harm individuals ultimately led participants to reject the option wherein both participants received no money.

The Fat Man thought experiment goes as follows: a train is headed directly toward five people chained to the tracks who cannot free themselves. You observe this from on top of a bridge under which there is a switch that would divert the train if a large weight were placed on it. Your own body weight is insufficient, but right near you on the bridge there is a heavy-set man whose weight would suffice to turn the switch. The only way of saving the five chained-up people is to shove the other person off the bridge and onto the switch that would divert the train onto another track. Should you push that person over? According to the serotonin experiments, individuals with high serotonin levels are less inclined to personally harm the fat man for the sake of saving the five people in chains.

Two Manchester-based bioethicists, John Harris und Sarah Chan, thus criticize the optimistic interpretation of the serotonin research as follows (Harris and Chan 2010):

1. Saving five people by sacrificing one could be thoroughly justifiable on moral grounds. They refer to the example of Jasper Schuringa, who stopped the terrorist Umar Abdul Mutallab from igniting a bomb on a flight to Detroit on December 26, 2009. It seems that he saved the lives of 290 people on that flight. If Schuringa's serotonin levels were higher, he may not have made the morally preferable decision or achieved the superior result.

2. If serotonin levels influence behavior, then this would simultaneously lead to the consequence that high serotonin levels reduce the
capacity to make moral judgments. In the case described above, this
would have had fatal consequences for nearly 300 people.

Crocket answers the second objection, however, by saying that she
does not share its premises. She furthermore addresses Harris and Chan's
judgment about the Fat Man experiment by pointing out that they have
a different understanding of harm than her. Her understanding of harm
applies only to cases where individuals personally and directly harm
other individuals through their own actions. Harm, or the loss of life,
through the avoidance of action does not count as harm in her sense of
the concept (Crockett 2010).

These examples, presented here in a general overview, show the variety, relevance, and timeliness of the problems discussed in this book.
These reflections give only a preliminary insight into the spectrum of
topics on the enhancement of emotions. Future, more speculative studies will certainly provoke new ethical challenges and raise additional
questions.

FINAL THOUGHTS

In this section it was not my concern to represent a certain moral attitude
regarding the different techniques for the enhancement of emotions and
morality, but rather to raise a series of questions that become relevant
in the context of this new field of research and that are also addressed
in the discourse of transhumanism. I hope that I have done justice to
the explosive and relevant nature of this research area and that further discussion will be initiated to help provide answers to the moral
and pragmatic challenges of these topics. I would like to end by briefly
presenting my own views on the use of enhancement technologies to
promote morality. In general, I consider the technological promotion
of characteristics and skills to be a promising research field, in which
numerous significant and worthwhile developments and advances have
been or can be expected. With regard to moral enhancement, on the

other hand, I would like to break my neutrality and express a few reservations, so that this overview does not go entirely without comment.

On the basis of the research results mentioned, I consider it quite obvious that basic moral attitudes can be influenced by neuroenhancers or other biotechnological measures. In contrast to many English-speaking researchers, such as Julian Savulescu, I consider moral enhancement to be the least relevant in the context of enhancement debates because the question of desirability and implementation does not suggest a simple answer. The situation is different from that of other enhancement technologies. Many people want to live healthier for longer, have higher intelligence, and be able to concentrate better. To be more moral may not be a core wish for some people. To put it bluntly, this may be a consequence of the fact that morality is generally in the interest of non-moral people, and that moral people are more likely to be exploited and oppressed. If this assessment is correct, then the question of motivation arises: Why fall back on technologies of moral enhancement? One possible answer is to escape legal sanctions. A convicted sex offender can be pharmacologically castrated in order to shorten his time in prison. But where should the motivation for taking these remedies lie unless someone is under threat of legal sanctions?

The background against which ethicists like Savulescu address the question of moral enhancement is the view that the global problems that could result in the extinction of humanity must be solved fast. Does this not mean, however, that these technologies must then also be used globally against the will of individuals? Such a measure would neither be easy to implement nor desirable, as it would undermine the cherished and fragile achievement of the norm of negative freedom that allows people to determine their own actions according to their own idiosyncratic notion of what they consider living well. I advocate strengthening this standard even further rather than restricting its effectiveness through global, paternalistic, coercive measures. Undermining this norm indicates totalitarian tendencies. Historical experiences with the Third Reich and other dictatorial regimes should have made it clear to us, however, that such movements must be avoided at all costs (Sorgner 2010b). However, if moral enhancement cannot and should not be imposed on people globally, there is still the possibility of the national

30

enforcement of moral enhancement as a coercive measure. Analogous to compulsory vaccination, which no longer exists in Germany but still exists in numerous other countries (e.g., the United States), a legal obligation for moral enhancement could also be adopted. Such a regulation may not be ruled out in practice, but in no way does it seem to me to be in the political interest of the country concerned since a country with a particularly moral population reduces its long-term chances of survival in the context of global competition.

For the various reasons I have just mentioned, reflecting on moral enhancement technologies seems less relevant to me than reflecting on other types of enhancement meant to promote human performance. However, the complexity of the enhancement debates can hardly be underestimated, and neither can their historical and systematic relevance. The enhancement of emotions and morality will certainly not be the last major enhancement debate.

CHAPTER 3

Pedigrees of Metahumanism, Posthumanism, and Transhumanism

Transhumanism is the most dangerous idea in the world. American political scientist and intellectual Francis Fukuyama holds this view at least.[1] That claim is true if one starts from a Christian or Kantian view of the human being. It is not yet widely known what transhumanism stands for specifically and which cultural, ethical, political, artistic, and philosophical views it implies.

Within this chapter I turn to a problem often encountered within contemporary schools of thought. When similar phenomena have names that are cognates of one another, they are often misleadingly conflated even if there are disparities in the concepts' meanings; for example, Habermas speaks of a "naturalistically inclined posthumanism" even when he is explicitly discussing transhumanism (Habermas 2001, 43).

This problem is especially rampant for three related philosophical approaches known as metahumanism, posthumanism, and transhumanism. All three cultural movements can be clearly distinguished from each other. This book is focused on explicating transhumanism. Since the other movements also come up now and then, and all of these approaches have currency today, I will at least sketch out the pedigrees of these pathways in the context of this text. By remarking on the different relationships that the major intellectual movements beyond humanism have with modernity, Enlightenment, criticism, reason, and the philosophy of history, I can better delineate the commonalities and differences between these current schools of thought.

COMMONALITIES AND DIFFERENCES

There are a great variety of thinkers, movements, and topics—both cultural and scientific—associated with debates about how to extend contemporary thinking "beyond humanism." Not all concepts in this domain already have clear, concise, and uniform meanings. The post-human of transhumanist discourses differs from the posthuman of posthumanist discourses. What the two concepts of the posthuman stand for exactly is just as unclear as the relationship between meta-, post-, and transhumanism. Some consider posthumanism to be a general term under which transhumanism can be subsumed. Others argue that these two concepts refer to two distinct social and cultural movements with different goals, origins, and genealogies. Within this section, I compile some initial considerations regarding the different pedigrees that can be identified with the possibility of transcending humanism.

The term "pedigree" fittingly encompasses central motifs of the debates not only because it refers to ancestors, one's own origins, and blood relations, but also since it is at home in the discipline of zoology, especially to describe an individual animal's lineage. By addressing this concept in the context of the posthuman or in the realm of discourses beyond humanism, a central element of these thematic contexts is emphasized: people are no longer regarded as the crowning glory of the natural world or as entities that categorically differ from nature in its purity, but are understood as only *gradually* differing from other natural beings. In contrast to the Christian and Kantian view of humanity, the different philosophies that go beyond humanism make the case for a new modesty about being human, because they no longer regard human beings as infinitely different and better than other natural beings.

Humanism is another concept with an enormous plurality of meanings. There is an ancient humanism, a Renaissance humanism, a Christian humanism, a Kantian humanism, and also a secular humanism. Within this book, I generally refer to a special version of humanism that embraces a categorically dualistic ontology of the world. In order to understand this issue, the origin of the word "humanism" must be analyzed briefly.

In 1808, the German theologian and philosopher Friedrich I. Niethammer coined the word "humanism," which occurred in the context of the examination of educational concepts (his book *Der Streit des Philanthropinismus und des Humanismus in der Theorie des Erziehungsunterricht unserer Zeit*). This intellectual context is not surprising, since the word "humanism" derives from the Latin word *humanitas* and in particular from Cicero's use of the word. This Latin word has a multitude of meanings but was identified in ancient times with the Greek term *paideia* (παιδεία). The identification of *humanitas* with *paideia* is implicit, for example, in Cicero's *De Oratore* (1.71). The two terms are then explicitly identified in the text of *Noctes Atticae* (13.17) by the second-century Latin author Aulus Gellius. Within this tradition, the word *humanitas* includes both logical and ethical elements, which is also true of the word *dignitas*, which Cicero was the first to attribute unambiguously to all human lives in his book, *De Officiis*. Cicero developed this use of the word *dignitas* from the term *humanitas*. Here dignity (*dignitas*) is used not only to highlight the distinction of certain individuals, a traditional use of the term that still occurs within Cicero' s work. Instead, dignity now applies to all people *qua* humans, or, *qua* rational beings, which are identical categories for Cicero. For Cicero, all people are rational and possess dignity, which is why they should also be treated with particular consideration by other people (Sorgner 2010b, 30–50).

Cicero's concept of dignity was later taken up by Kant and is especially influential in the present, considering that Kant's concept of dignity inspired the concept of dignity in the German Constitution. This term is an ontological one since it implies a certain anthropology. However, it is also an ethical one since the respective entities should also be treated ethically in a particularly good way. This diversity of meaning is similarly contained in the concept of *humanitas*. On the one hand, it implies a categorically dualistic ontology that entails a categorically special position for humanity. The term *paideia* has analogous connotations, which can be seen in Plato's analogy of the divided line in the *Republic*. Cicero's case for the immortal soul illustrates parallel structures with regard to his concept of *humanitas*. Such dualistic views are firmly rejected within secular, naturalistic, atheistic, or evolutionary variations of humanism, which is why it can be doubted whether they are aptly named forms of humanism. However, there is still an ethical dimension to *humanitas* that has traditionally suggested the goal of self-perfection. This goal is evident in Plato's characterization of the philosopher king and in the description of the Stoic wise man, who is always virtuous. Although a strong concept of ethical perfection is not necessarily associated with transhumanism, such an orientation can be described as the predominant concept of good within transhumanism. I will later speak out in favor of a weaker idea of the good—namely, the radical plurality of goodness, which can also function in accordance with transhumanism. At this point, though, it is crucial for me to point out that transhumanism does indeed go beyond humanism and, like secular, naturalistic, or evolutionary humanism, is not really in the tradition of humanism if the etymology of that word is still relevant to its current meaning.

The dualistic concept of humanism is rejected within the majority of positions beyond humanism, but not for all; for example, there are Mormon transhumanists who (mostly) start from the existence of a metaphysical world, that is, from the existence of a transcendent world that categorically differs from our sensual world with regard to its ontological characteristics and traits. But here I am just trying to present a general overview of the major debates, which impels me to overstate and overgeneralize some phenomena. This way I can point out relationships and tendencies that would otherwise remain obscure.

TRANSHUMANISM

The majority of transhumanists purport a mutual, materialistic, naturalistic, relational, or immanent understanding of the world. The theory of evolution plays a central role in their understanding of humanity. Transhumanists take evolution seriously and therefore strongly believe that everything that has evolved to its current form is likely to evolve yet again in the future if it does not die out first. Further developments in human life could also lead to our extinction. Transhumanists wish to implement the latest technologies as a way of promoting human survival and prosperity. This basic attitude can serve as the basis for a general definition of transhumanism: *transhumanism embraces the use of technologies to increase the likelihood that posthumans may emerge*. It is this attitude that is characteristic of all transhumanists. Transhumanists do differ significantly in their understanding of the posthuman and the most promising enhancement technologies (*Techniken*). The posthuman can designate a new species apart from the human. In this case, the transhuman is another central concept that represents a member of the human species in the process of becoming a posthuman (Bostrom 2005a, 11). It is also possible, however, that the posthuman would be a member of the human species that already has at least one trait that is beyond those that currently exist (Bostrom 2009). Another option is that the posthuman is no longer a biological entity but exists in digital cyberspace (More 2013, 7). In this case, the cyborg could be the most promising way to create the posthuman, as cybernetic organisms belong to both the organic world and the realm of digital and mechanical technologies, making them ideally suited to act together as mediators between present human beings and the digital posthumans of the future.

It is also an open question which technologies are most promising for increasing the likelihood of posthuman emergence. The following enhancement processes are at the center of the various debates on these issues:

1. genetic enhancement, for example, genetic enhancement through selection or modification
2. morphological enhancement, for example, morphological enhancement through cosmetic surgery
3. pharmacological enhancement, for example, through Ritalin or modafinil
4. cyborg enhancement, for example, through deep brain stimulation (various companies may already be working on some kind of BrainStation or iBrain)

It is also an open question which characteristics are decisive for an entity to be properly called posthuman. Some possibilities are as follows:

1. superintelligence
2. a strong memory
3. a long healthspan

4. a strong physiology owing to physical strength, beauty, or health

5. morality

Because of the variety among the concepts of the posthuman, it is important to distinguish two main strands of humanism. Of importance here is the fact that transhumanists not only count on the latest technologies as means to achieve their goals, they also expect new technologies to support and complement existing technologies. Pharmaceutical products could be helpful in reaching certain states of consciousness (e.g., mindfulness).[2] This does not mean that transhumanists reject traditional meditation techniques as ways of achieving mental equilibrium. It may well be the case that the new technologies can complement, support, and also improve extant techniques to achieve the same goals. Both options can be helpful in achieving specific goals. Genetic enhancement through modification could become an important technique to increase the likelihood of posthuman development, which does not mean that education would necessarily become irrelevant. Techniques and technologies (*Techniken*) can both be means to change a human genome (Sorgner 2013a). These reflections make clear the first central argument regarding human evolution and development toward the posthuman: through the genetic change a development toward the posthuman can be driven forward, so that the posthuman will either be a member of a new species or at least possess a characteristic that lies beyond the characteristics of the people currently living (e.g., the ability to perceive UV rays). This understanding of the posthuman may imply that he is necessarily radically different from the people living today. Yet this use of the word "posthuman" may not differ much from the use of the same word among posthumanists. I will discuss their use of the term separately in the following section.

On the basis of this understanding of human development, it can be presumed that the use of the latest technologies (*Technologien*) and the associated possibilities to promote human change are not categorically different from the techniques (*Techniken*) already known to us. After all, people continue to do what they have always done: they invent and use techniques to make our lives easier, more fulfilled, and better. This kind of transhumanist thinking does not necessarily mean that transhumanist attitudes are aimed at a radical break with traditionally human habits. Transhumanism understood in this way would simply mean pushing the use of new techniques while promoting goals that were already particularly relevant in most cultures, for example, cognitive abilities, intelligence, and healthspan.

There is another way of surpassing humanity that is often discussed among transhumanists. It puts the emphasis on the possibility that the posthuman is a cyberspace entity.

In contrast to carbon-based transhumanism, silicon-based transhumanism focuses in particular on the technology of downloading the personality onto a hard disk so that it can be multiplied, integrated into a new organism, or could continue to live alone in cyberspace. These processes are called "mind uploading" or "whole brain emulation." In contrast to carbon-based transhumanism, silicon-based transhumanism seems to be based on a different anthropology. The first suggests a naturalistic anthropology. The second version seems to be based on a kind of dualism—so runs the accusation of numerous posthumanists and many other critics of transhumanism, in any case.

Most transhumanists argue for a naturalistic view of humanity, and they therefore propose that a detailed brain scan could be used to develop a software model that could run on different hardware models. This basic attitude in turn seems to be based on a dualistic understanding of the human, according to which the mind exists separately from the body, so that the mind can also be separated from the body. If both judgments are correct, it is obvious to ask how this stance is not self-contradictory. Naturalism and substance dualism are mutually exclusive philosophical positions. However, the charge that the transhumanists who affirm mind uploading support substance dualism is incorrect. The software-hardware divide among transhumanists also takes place within a naturalist anthropology, which does not have to be described as a possibility of thinking. I for one do not rule out the possibility of mind uploading. But it would probably only be possible given if the functionalist theory of mind were correct. Only in this case could there be a consistent embrace of a temporal anthropology and the possibility of downloading the personality to a hard disk (More 2013, 7). Anyway, this specific question is a fascinating area of research, and further work is necessary to be able to approach the central research relevant to this.[3]

From my point of view, however, a carbon-based transhumanism and the resulting understanding of the posthuman is more consistent and probable with regard to the possibilities of human development. Moreover, this understanding of transhumanism has many similarities and overlaps with philosophical posthumanism.

POSTHUMANISM

The concept of the posthuman also plays an important role in posthumanism. However, the meaning of this word differs in its use by posthumanists and by transhumanists. Among transhumanists, the term refers to an advanced human being. Among posthumanists, however, the concept of the posthuman stands for a new understanding of humanity, whereby posthumanists also possess a special methodology and take up a continental European style of philosophizing as a way to formulate this new anthropology (Braidotti 2013; Hayles 1999). Considering their special angle on the topic enables us to formulate a fundamental definition of posthumanism: *posthumanism represents the attempt to resolve categorical dualities within statements based on considerations of the relationship between technology and human beings, since it seems plausible that posthumanism is the most apt way to grasp the completeness of the world in linguistic form.* This basic position has many implications. With regard to the concept of the posthuman, there is a need to break away from an anthropology that sees the human being as a being consisting of an immaterial soul and a material body. For this reason, posthumanists propose a nondualistic view of human beings. The concept of the human in humanism implies that human beings have a material body and an immaterial body, that is, a soul or spirit. The posthumanist attempt to transcend categorical dualities goes beyond this concern and also suggests a revision of the traditional relationship between humans and animals, humans and digital and mechanical machines, and between nature and culture. The attempt to transcend categorical dualities is only one aspect of posthumanism.

When we talk about posthumanism trying to move away from a dualistic version of anthropology, which implies that human beings consist of a material body and an immaterial soul, then this discourse implies that human beings have always been posthuman (Hayles 1999, 291). Based on this understanding of the posthuman, it would be pointless to ask which moments or processes in time gave rise to the human's phylogenetic development into a posthuman. To have always been posthuman requires that humans have always been part of gradual evolutionary processes that have allowed us to develop from our common ancestors, the great apes, to the people we are today. As posthumans we have always been dependent on technology, and there is no clear categorical distinction between nature and culture, body and soul, or genetic and environmental influences. The posthumanist attempt to break away from dualistic anthropologies is based on the plausibility of the assumption that we see ourselves as nondualistic entities. However, this view had not prevailed during the Enlightenment. At that time, it was widely accepted that humans possess a material body and an immaterial soul, Descartes and Kant being the leading philosophical representatives of this human self-image. To see ourselves as posthumans does not imply a phylogenetic development from human to posthuman, but it demands that we were always posthumans with regard to our anthropology. However, a cultural development has taken place with regard to our anthropological self-image. During the Enlightenment we generally saw ourselves as beings with a material body and an immaterial soul, but nowadays we increasingly see ourselves as nondualistic entities. This shift in our self-image has decisive implications, especially with regard to numerous ethical and legal issues (Sorgner 2010a; 2013b).

Another important theme of posthumanism concerns the concept of truth. Posthumanist thinkers believe that we cannot expect to arrive at truth that corresponds with reality since this goal cannot be achieved in a plausible way. The pedigree of posthumanism is directly related to postmodern thinking, which is why posthumanists as well as postmodernists consider it appropriate that only plausibility can be an objective of philosophical investigations (Sorgner 2013b, 135–44). Each judgment, in principle, represents only an interpretation, whereby the concept of interpretation does not imply that the part presented must be wrong. All that the concept of interpretation demands is that any judgment *can* be wrong. Since no theory has so far revealed itself as the only true one, we

are not in a position to distinguish what is true and what is a false interpretation. It could be the case that we will never be able to make such
a distinction. As long as no theory has revealed itself as the only true
one, postmodern perspectivism is to be regarded as the most plausible
theory of cognition, although it cannot guarantee reliable knowledge
in correspondence with the world. However, this assessment does not
mean that it is impossible to distinguish between plausible and implausible judgments on the basis of currently dominant parameters, which
are subject to change depending on the zeitgeist.

BETWEEN POSTHUMANISM AND TRANSHUMANISM: METAHUMANISM

This brief description of posthumanism and transhumanism shows
that both move beyond traditional humanistic anthropology, that is,
beyond any theory that considers humans to consist of material and
immaterial aspects.

This view is of central relevance to all representatives of both movements, and it is this insight that both movements share. Nevertheless,
a certain hostility between the members of both cultural movements is
widespread and must be acknowledged. The main reasons for this hostility are that the movements differ in the following points:

1. With regard to the use of language, style, and methodology, the
   writings of the representatives of both movements differ radically.
   Transhumanists have a linear way of thinking, use technical terms,
   and mostly fall back on a scientific methodology. Posthumanists, on
   the other hand, have a nonlinear way of thinking, use metaphorical terms, and have a hermeneutic methodology.
2. The origin and genealogy of the two movements are radically different. Transhumanists are deeply rooted in the English tradition,
   which is closely linked to Darwin's theory of evolution and Mill's
   utilitarianism. The intellectual founder of transhumanism, Julian
   Huxley, fully exemplifies this tradition. Posthumanists are part of the
   continental philosophical tradition and therefore closely connected

with literary theory and cultural studies. They also usually take narrative and radical-pluralistic approaches to ethical questions.

The above-mentioned anthropological differences already indicate that both movements may have more in common with each other than they are often willing to recognize. An alternative approach, called metahumanism, has recently been established. It champions weak versions of posthumanism and transhumanism, strives to establish a relation and dialogue between both discourses, and at the same time represents an alternative to them (del Val and Sorgner 2011). It is described as metahumanism because its thinking surpasses a dualistic conception of humanism (meta can mean "beyond"), but it also occupies a position between posthumanism and transhumanism (meta can also mean "in the middle of"). Metahumanism can be understood as follows: *metahumanism strives to mediate among the most diverse philosophical discourses in the interest of letting the appropriate meaning of relationality, perspective, and radical plurality emerge.* One suggestion from this philosophical branch was the introduction of the word "metahuman" to refer to human beings as worldly, nondualistic, and relational. Posthumanists usually have a materialistic image of human beings. The term "metahuman" is thus a further development of the posthumanist concept of the posthuman, whereby it represents a relational and not a materialistic image of human beings, which implies a great proximity to neo-Spinozist thought. Because of its prefix, this term cannot be confused with the transhumanist terms "transhuman" and "posthuman," which represent different levels of further developed human beings.

All three terms can thus be used within one discourse, representing different facets of one new understanding of humanity, which holds that we humans are also part of the evolutionary chain and will continue to develop if we do not go extinct beforehand. In contrast to the dominant transhumanist concepts of perfection, however, metahumanism represents the radical plurality of good, which implies that a nonformal conception of good must always be implausible.

Without being able to give a detailed description of the historical developments of the different movements,[4] I will try at least to sketch a general map of the movements that surpass humanism in scope. In particular, the relationship of the three movements to modernity and postmodernism will be examined in more detail.

MODERNITY AND POSTMODERNITY

It is still an open question, during which period of time humanism dominated our thoughts and actions. By analyzing the concept of humanism more closely, the meaning of the various movements beyond humanism is explained more explicitly yet. Ihab Hassan already stated in 1977 that five hundred years of humanism could have come to an end (Hassan 1977, 843). From Sloterdijk's point of view, humanism has dominated Western cultures since the age of Stoicism (Sloterdijk 2001, 304). Considering that Stoic philosophy exhibits many dualistic tendencies, it is certainly arguable that humanism has dominated Western cultures at least since that time (Sorgner 2010, 38–40). I disagree with Sloterdijk's history, however, since humanism must be regarded as a central cultural force within Western cultures going back to the time of Plato. This is evident in the myth of Er from book 10 of Plato's *Republic*, which illustrates the core, strongly dualistic implications of Plato's philosophy. The three proposals concerning the date when humanism was born have significant philosophical implications because different dualities are linked to the different concepts.

In Hassan's concept of humanism, the birth of humanism goes hand in hand with Descartes's philosophy, which implies that the relevant dualism identifies the human being with an immaterial soul and a material body. Animals, plants, and stones all fall into the same category. They all belong to the material world alone.

In Sloterdijk's conception of humanism, the birth of humanism is connected with Stoic philosophy; then the dualism in question would be of another order. The Stoics regard animals as animate beings. However, the animal soul is of a different kind than the human soul since only humans, including slaves and women, possess a rational—that is, a reasonable—soul. The rational soul enables people to think and speak. Animals, on the other hand, do not possess these abilities according to the Stoic perspective.

In my own understanding of humanism, the philosophically and culturally decisive step towards humanism took place with the myth of Er in Plato's *Republic*, in which the immortal soul of the warrior Er leaves his body upon his death, but then returns again, even if elements of dualistic thinking can even be traced back to Zoroaster, whose religion introduced a rigid good-evil dichotomy. A decisive difference between Platonic humanism and that of the Stoic philosophers is that the Stoa demands a minimal equality of all rational beings solely on the basis of human rationality, whereas Plato's philosophy holds that not all humans are able to use their rationality to an equal degree since this ability depends on what kind of soul one possesses, a soul of gold, of silver, or of iron (Plato's so-called noble lie). Another important line of development within humanism can be found in the birth of the ancient drama, which emerged not long before Plato wrote, and which will be addressed in more detail below.

Regardless of whether Hassan's, Sloterdijk's, or my own view on the birth of humanism is correct, it is important to note that many researchers and scholars today agree that "a crisis of humanism" is taking place across discourses around the world, as already known by Badmington (Badmington 2000, 9). The terms post-, trans-, and metahumanism thus stand not only for cultural movements and a special way of thinking, but also for a description of a certain epoch—namely, the period of time that follows humanism. In contrast to the concepts of modernity and postmodernity, the concept of humanism was not widely used to refer to a certain period of time. Rather, the usual use of the word humanism denotes a certain attitude and a certain way of thinking about the world. In order to strengthen the use of this concept as an epochal designation, the following section will focus on the established cultural periodization of modernity and postmodernity. In this context I will refer to distinctions proposed by Marquard (1995, 92–107), which represent a classification of individual concepts that are widespread but inappropriate.

From Marquard's perspective, we still live within modernity today, as characterized by a pervasive affirmative attitude to progress (Marquard 2000, 50; 1989, 7). In his view, modernity can be replaced by one of the following three ways of thinking:

1. A pro-modern variant of the philosophy of history: thinkers who hold such an opinion often talk about the dark Middle Ages, which are overcome by the Modern Age, which in turn is dominated by an affirmative attitude toward progress. Marquard considers Hegel to be a thinker who distinctly personifies this understanding.

2. An antimodern variant of the philosophy of history oriented toward the past: this attitude is characterized by the fact that it regards a premodern epoch as a cultural ideal, which is why any progress must be regarded as a descent by the views prevailing at that time. Novalis with regard to the Middle Ages, Winkelmann with regard to antiquity, and Rousseau with regard to a precultural nature represent three variants of such an understanding of history.

3. An antimodern philosophy of history oriented toward the future: here, too, the Enlightenment is regarded as decline and decay, but at the same time a perfect future is expected, which cannot be actively worked toward or which fatefully awaits us. Marx's thinking is a particular version of this philosophy of history. His ideal future society represents a classless society that lies ahead as a goal for further developments (Marquard 1995, 95–97).

Marquard's way of conceptualizing history is helpful, but it is not comprehensive. Neither post- nor transhumanism fit in well with Marquard's proposed schemes. The same applies to postmodernism, and posthumanism can be seen as a further development of postmodernism. Postmodern philosophers, such as Derrida, Foucault, or Lyotard, are neither pro-modern, nor do they necessarily see the Enlightenment as a decline, but they also do not project a desirable future ideal. They do not even propose that the Enlightenment tradition of criticism must be continued. With regard to historical progress, a certain kind of indifference can be analyzed among these thinkers. This is one reason to doubt that modernity can be reduced to the affirmative attitude to progress proposed by Marquard. In the following section, I will propose a more complex way of thinking about modernity so that the various attempts at thinking beyond humanism can be integrated into it.

Marquard is certainly correct to call attention to the fact that an affirmative attitude to progress is a fundamental aspect of modernity.

By emphasizing this point, he also draws a clear distinction between modernity and the Middle Ages. During the Middle Ages, Christians believed they possessed the one and only truth in correspondence with the world. It is therefore also clear that progress was of no value in such an approach. The reasons for that devaluation are just as clear. If one is aware of the one and only one unchanging truth, then every development must represent a movement away from truth. If the culture of the Middle Ages considered itself the manifestation of this truth, any cultural development away from this structure must represent a descent or development away from God, truth, and the good life. This development began when the value of critique was first recognized. All those central premises that were taken for granted during the Middle Ages were criticized in modernity. Such criteria were applied at different levels and in different areas and disciplines. Sloterdijk's *Critique of Cynical Reason* (1983) provides an excellent overview of the most important critical traditions during this epoch. In the chapter "Critique of Revelation," for example, he explains that biblical revelation is not justified by reason.

Sloterdijk's book presents eight persuasive "unmaskings" of medieval thought: in the "Critique of Religious Illusion," he argues that religious institutions were assigned the task of addressing people's fear of living their own lives; the "Critique of the Metaphysical Illusion" notes that metaphysical questions can be asked but not answered; the "Critique of the Idealistic Superstructure," on the basis of which it can be doubted that the political order is a just one; or the "Critique of Moral Illusion," in the context of which he criticizes the "lustful monk, the bellicose prelate, the cynical cardinal, and the corrupt pope" in great complexity and detail (Sloterdijk 1988, 42).

The examples mentioned here make it seem plausible that in addition to progress, criticism, and reason, other concepts are closely linked to modernity. Modernity made progress because reason was used to criticize absolute attitudes that were traditionally prevalent. Many consider Descartes's "cogito ergo sum" as the beginning of modernity because his skeptical method encompasses all central concepts of modernity: the unified subject uses reason to critique the limits of knowledge.

Not only in the field of philosophy did important developments take place during the Enlightenment, but also in the fields of society, politics, and the natural sciences. Both Nietzsche and Heidegger aptly pointed out that the importance of the natural sciences increased steadily during this epoch. As a result, the various irritations Freud spoke of have also occurred: First, the Copernican irritation, which led to the realization that it is not the Earth but the Sun that is at the center of our solar system. The second irritation followed with Darwin's theory of evolution and the realization that apes and humans have common biological roots. The third irritation was inflicted on us by Freud himself. I will take a closer look at this point a little later, as it also entails a movement that leads away from modernity and transcends it.

In medieval politics, the aristocracy represented the divine order. Their importance was gradually replaced by liberal and democratic systems in which tolerance, autonomy, and negative freedom play a central role. Freedom includes that one must not limit the freedom of others and that one must not harm other people. A strong and detailed concept of good that prevailed in medieval politics was slowly but steadily replaced by a pluralistic understanding of good.

A development closely related to these changes can be noted within the arts. In the Modern Age, a development occurred that separated the population from the art world. As long as principles of faith were shared, it was possible for members of all areas of the congregation to have a connection to the arts. However, during the process of creating a separate artistic world, artists themselves gained importance, and the arts no longer had the obligation to serve one specific role, such as praising God, glorifying the political order, and telling religious stories. Since 1800, the artist, often seen as a genius, was regarded as a specific subject who created his works of art out of an inner necessity (Sorgner and Fürbeth 2010).

Once again, some central characteristics and features of modernity can be worked out on the basis of these thoughts. Faith in the value of progress came about with the help of reason. Faith in progress is closely linked to the concept of the Enlightenment. Before the Enlightenment, truth was eternal and unchangeable and accessible with the help of reason. During the Enlightenment, the meaning of reason changed, since various forms of rationality were finally acknowledged. Reason became diverse. In this way, reason developed the potential to question its own meaning. An evolutionary reason can no longer guarantee the unity of reason. The importance of reason has been undermined by an increase in the relevance of the sensual world and empirical data, as well as by the growth in the importance of science and technology. Both developments played an important role, especially in the English-speaking world. At the same time, reason was still considered important in order to grasp the truth. It was also closely linked to the critical tradition that plays a particularly important role within continental philosophy. Descartes represents a paradigmatic case for this basic philosophical understanding. Reason can be found in every human being—namely, in the part of a human being that represents the unified subject. By referring to this aspect of reason, it was possible to criticize old prejudices, absolute worldviews, and religious hierarchies. In this way, the importance of the metaphysical world diminished, and the importance of the sensual world increased, thus again promoting the relevance of science and technology.

The medieval world attributes everything absolute and perfect to God. During modernity, human subjectivity gradually attracted more interest than divine subjectivity, while subjectivity was still closely linked to rationality. To make reason a new universal foundation fomented new tensions because the reliance on individually executed judgments undermined the possibility of a universal foundation for reason. The rational subject uses reason to criticize judgments. Progress has been made, however, through this line of thinking. Because this is a movement initiated by the subject, and the movement is also supposed be in the subject's interest, progress is closely linked to the belief that the general conditions of human life can be enhanced.

However, we have already recognized that there is a potential tension within the Enlightenment project that necessarily led to a self-undermining process. Process thinking was responsible for Darwin developing the theory of evolution. By taking physical processes and the secular world seriously, a process was initiated that also applied this self-understanding to the areas of knowledge. If, in turn, reason alone is a skill that was created on the basis of evolutionary processes and is therefore something that is not present in all men in an identical manner (in the strictest Leibnizian sense), then there is no longer any reason to assume that reason lets us recognize the truth, at least not in the sense of the correspondence theory of truth. It is also no longer possible to posit a unified reason that represents our unchanging inner human nature. The first insight leads to doubt that the truth can be grasped as correspondence to effects. This approach is astutely described in Nietzsche's perspectivism. The second insight implies the fragmentation of the formerly unified subject and is clearly evident in Freud's psychoanalytic theory. Both views are central concepts bound up in postmodernism.

Since both of these concepts play a central and significant role in Nietzsche's philosophy, it is no surprise that many philosophers regard him as an architect of postmodernism. Habermas, Vattimo, Sloterdijk, Derrida, and Foucault agree on this (Habermas 1990, 83–105; Vattimo 1988, 164; Sloterdijk 1987, 55; Robinson 1999, 34). Nietzsche's perspectivism in particular, according to which every perspective represents an interpretation, is important for this role to come to fruition (Sorgner 2007). Most continental philosophers after Nietzsche have continued to work on the basis of this insight. The members of the Frankfurt School are an exception and do not resolutely agree with this attitude since they still think in terms of a unified rational subject. Besides them, a multitude of perspectival approaches can be seen within contemporary continental philosophy, which is why it is possible to speak of a long-lasting predominance of postmodernism within this tradition.[5] The hermeneutic tradition can also be regarded as an approach that is closely connected with postmodernism, which has already been recognized by Marquard (Marquard 1981, 20). According to Rorty, truths are expressed in sentences, and sentences are created by humans (Horster 1991, 88). In Derrida's view, all language is metaphorical, which implies that he doubts the possibility of linguistically expressible knowledge of truth (Robinson 1999, 38). It was Nietzsche, once again, who had already noticed both of those points and provided an explanation for why language consists exclusively of metaphors and relations (Nietzsche 1988, 1:879). Postmodernism posits a close connection between reason, language, and doubt about the possibility of knowing truth. Reason for its part is closely linked to the ability to formulate linguistic judgments. Linguistic judgments consist of metaphors and relations. How could it still be possible then for linguistic judgments to be more than a collection of metaphors and relations?

Without a unified rational self that enables us to recognize truth in correspondence with reality, the concept of spirit must also be rethought. Spirit is no longer something eternal and unchangeable, but something that has arisen on the basis of evolutionary processes. Humanity could once claim ontological unity because human beings were understood as united, rational entities, but humanity is no longer unified in the same way. As a result, the human became a fragmented entity, and once again it was Nietzsche who already articulated this insight (Nietzsche 1988, 11:650). This insight was further developed in a more scientific and detailed way by Freud, who was responsible for the third offense against human nature. This offense comes from the fact that the rational subject is no longer the master in his own house.[6] Habermas too presents Nietzsche as a turning point on the way to the postmodern era and referred to two criteria that mark postmodernism: 1. the dissolution of subject-object distinction; 2. the doubt as to the possibility of being able to know the truth (Habermas 1990, 96–97). Habermas is critical of these developments, as they are represented by thinkers such as Derrida, Foucault, and Bataille, and by referring to these thinkers as young conservatives, he expresses his dismissive attitude toward this basic attitude. Habermas had unstated reasons for selecting this language, and these reasons emerge clearly through an example mentioned by Nussbaum, which I will discuss in detail later in this chapter (Nussbaum 1995, 64–65). Habermas and Nussbaum are critically opposed to postmodern thinking. The question addressed here is indicative of the relationship between postmodern philosophers and those with a universalist understanding of the anthropology of rationality. Habermas's accusation of "conservativism" comes off unconvincingly, because the term "conservative" refers above all to the intention to support and maintain established structures. Postmodernism, however, is definitively keen on going beyond modernity. Habermas's philosophy remains within the framework of the paradigms of modernity, which makes it more plausible to describe him as a conservative thinker.

Another consequence of postmodern thinking is the increasing importance given to the concept of games. The language game in Wittgenstein's later philosophy can also be regarded as an expression of postmodernism. The same applies to Feyerabend's slogan "anything goes," Duchamp's readymades, John Cage's 4:33, Dadaism, and Warhol's attitude that everyone could be famous for fifteen minutes. Everything becomes a game when there is no ultimate criterion for truth. Is this indeed the case? Does postmodernism itself not present a self-contradictory concept since it embraces perspectivism as well as the anthropological judgment that people are fragmented? Postmodernism therefore makes both an anthropological judgment and an epistemological statement, which seems self-contradictory. But is it perhaps not a problem for the postmodern to contradict itself if there is no ultimate criterion for plausibility?

MOVEMENTS BEYOND HUMANISM

Now we can finally deal with the various schools of thought that aim to surpass any form of humanism with a basis in a radically dualistic ontology. How do posthumanism and transhumanism fit into these above-mentioned general schemes? Posthumanism embraces both perspectivism and a secular, and therefore also fragmentary, understanding of humanity. On this basis alone, it is clear that posthumanists represent a further development of postmodernism. This assessment applies especially to postmodern perspectives such as those of Deleuze and Foucault, but not to those of Levinas and Derrida. Derrida doubts that the truth can be grasped in linguistic form, but nevertheless he posits a concept of truth as event, which disappears once we try to put that experience into words (Derrida 1991). Here it becomes clear that Derrida does not subscribe to an immanent worldview but is far more at home in the history of theology and of Jewish ethics. The main difference between posthumanists and postmodernists like Deleuze and Foucault is their worldliness (*Diesseitigkeit*), which is emphasized more strongly by posthumanists than by postmodernists.

Transhumanists also have a worldly or naturalistic understanding of humanity. But they also emphasize the importance of reason and a form of truth. In this way, transhumanism sees itself as belonging to the tradition of the Enlightenment (More 2013, 10). Is this an appropriate, adequate, and consistent way of reflecting on oneself, taking into account the considerations of the preceding pages? If it is the case that reason has an evolutionary origin, that is, that the ability of reason has arisen for 51 pragmatic reasons, then it can be doubted that reason can actually provide us with the one and only truth in correspondence with the world. However, this does not mean that reason should then be rejected or that it no longer has any meaning. Instead, it follows from this insight that the limits of the possibility of reason must be taken seriously: reason helps us pragmatically, but it is not able to communicate the truth to us in correspondence with reality. In any case, it must be emphasized that the concept of reason by modern thinkers such as Descartes and Kant should not be confused with the concept of reason within transhumanism; the two concepts are incompatible. Descartes and Kant argue that people possess a monistic, rational part and a worldly, material part. Transhumanists embrace a scientific, naturalistic, and worldly understanding of the human, and therefore they must reject the modern understanding of the human as paradigmatically represented by Descartes and Kant since an ultimately naturalistic anthropology does not provide for the possibility of the existence of nonmaterial reason.

However, a naturalistic anthropology does not have to be materialistic. Spinozism, relativism, and nondualism also represent possible variants of naturalism. For this reason, there is a clear gap between the rational concepts of the Enlightenment and those of the transhumanists (Hughes 2010).

Does this assessment mean that transhumanists must give up their affirmative attitude toward reason? I take it that these considerations should at least lead to the conclusion that the relationship between transhumanism and the Enlightenment is not as tense as it often sounds from the transhumanist self-understanding, and that the self-assessment of this relationship can be revised in part. Transhumanists take on a naturalistic view of reason, which implies that reason has much more limited capacities than the concept of reason within the Enlightenment. Applying reason can still be regarded as useful, but not because it is able to bring us into contact with truth in correspondence with the world, but rather for pragmatic reasons—namely, because the use of reason helps us in coping with our worldly challenges. It can therefore also be said that posthumanism and transhumanism share many more characteristics than is generally assumed, especially if we restrict our focus to carbon-based transhumanism.

Let us now turn to posthumanism. In this context, it is worth taking a closer look at an example mentioned by Nussbaum. She described the following situation: Western scientists introduced smallpox vaccination to India, threatening the Indian cult of Sittala Devi, a goddess to whom local people prayed in the hope of being cured of smallpox. Nussbaum attended a conference at which this situation was discussed and at which an elegant French anthropologist, decisively influenced by postmodern thinking, analyzed these processes in her reflections. The anthropologist emphasized that there was no privileged perspective on health, as had already been shown by Foucault and Derrida in an apt and convincing way. In such a process, the Western scientist would not adequately respect Derridean *différance* and would treat the locally dominant cultures in a degrading and colonial way. After her speech, one participant asked if it would not be better to live than to die, to which the anthropologist replied that this statement presupposes a Western essentialist concept of medicine based on dualistic opposites and categorically differentiating between life and death. But once one has freed oneself from this way of thinking, then it becomes possible to understand its difference from the Indian tradition (Nussbaum 1995, 64–65).

As the above example shows, many transhumanists and other academic philosophers see the postmodern way of thinking as problematic and doubt the intellectual value of even entering into public debates about such ways of thinking. I certainly share doubts about the intellectual value of the above-mentioned French anthropologist. I take it as a given the French anthropologist has not applied postmodern thinking in the appropriate manner. The problem raised here is not related to conclusive philosophical arguments concerning ontological issues. It is indeed unclear whether a Western dualistic or an Eastern nondualistic thinking would provide a conclusive answer to the ontological questions about humanity. However, the question of whether to give smallpox vaccinations occurs at a different level of inquiry. Whether or not a treatment is successful is not a question of the ultimate truth about the world; rather, this question has to do with what can be realized in this world. If it is the case that a smallpox vaccination in most cases immunizes people against smallpox, but prayers to Sittala Devi usually do not cure patients of smallpox disease, then we have a good reason to rely on Western medicine and smallpox vaccination instead of on the prayer to the goddess if one has the intention not to get smallpox or to be healed of infection. In these cases, experience has shown that Western medicine is more successful than prayers to the goddess. Relying on what has been successful in most cases does not mean that this method gives us any better understanding of the world. Relying on empirically established methods simply makes pragmatic sense. When postmodern and posthumanist thinking are applied to inappropriate topics, they could bring needless suffering on people who live traditionally. Posthumanists would benefit from approaching science with pragmatic ethics—like transhumanists do.

CONCLUSION

I have tried to show that there are elements of transhumanism that can be relevant to posthumanism, just as there are aspects of posthumanism from which transhumanism can benefit. If the exchange between both traditions were promoted, it would be possible to realize a complex thinking beyond humanism, as well as beyond the prevailing local cultures, since posthumanism and transhumanism are linked to different local cultures: transhumanism is linked to Anglo-American scientific culture, and posthumanism to continental literary culture.

How to perform metahumanist thinking: metahumanist thinking lies beyond a dualistic understanding of humanism, just as it lies between posthumanism and transhumanism. "Meta" means both "beyond" and "between." Metahumanist thinking strives to think beyond an understanding of the world that categorizes ontological dualities, such as the distinction between subject and object and between immaterial mind and material body. Metahumanism also moves between posthumanism and transhumanism, however, by addressing posthumanist and transhumanist challenges using a hermeneutic method inspired by Vattimo's weak thinking, his *pensiero debole*,[7] and applies it to the current challenges of the latest technologies (Sorgner 2013b, 135–44). In this way, the complexity of a Nietzschean and hermeneutic philosophy is taken seriously as well as the diversity and accuracy of the analytical ethical discourses. Metahumanism can be understood as embracing both a weak version of transhumanism and a weak version of posthumanism. At this point, at least a few selected differences between a strong and a weak version of both understandings must be outlined.

Transhumanists defend technologies that increase the likelihood that posthumans will emerge. A strong version of transhumanism posits that there are moral but not legal obligations to use certain enhancement techniques. Bioliberal Julian Savulescu, who is not a transhumanist but whose thinking closely resembles transhumanism, represents such a perspective in an exemplary way (Savulescu 2001; Savulescu and Kahane 2009). Another strong version of transhumanism argues that enhancement techniques necessarily promote the good life and that this insight must bear legal consequences. Aubrey de Grey, for example, holds this view (de Grey and Rae 2007, 335–39). A weaker understanding of transhumanism posits that enhancement techniques only promote the likelihood of many people leading a good life without necessarily requiring that transhumanist insights entail legal and moral obligations, which does not necessarily mean that such a position would have no implications for moral and legal rights (Sorgner 2013a, 2013b).

Posthumanism, in turn, is concerned with promoting nondualistic thought and action. A strong version of posthumanism would argue that humans differ only by degrees from other natural beings and that this insight should have legally binding consequences. Peter Singer is a representative of such a view (Singer 2002; 2009). A weaker version of posthumanism would argue that humans may indeed be distinguishable from other natural beings only by degrees, but it would not require that this insight result in legal reform. Instead, such a weak version would regard this position as no more than a legally legitimate perspective; that is, the law should not be forbidden from thinking and acting according to this insight (Sorgner 2013b).

The reflections on meta-, post-, and transhumanism presented here have at any rate shown that Marquard's attempt to philosophically describe modernity and the movements that follow it does not represent a comprehensive description of the present phenomena. Transhumanism at least seems to meet the requirements of Marquard's first group since it demands a pro-modern attitude to progress. However, transhumanism is not pro-modern in terms of its anthropology since modern anthropologies are based on a dualistic concept of the human, according to which human beings have an immaterial mind and a material body. This view is aptly exemplified by Descartes and Kant, but it is not shared by transhumanists. This is why it is problematic to classify transhumanists as entirely pro-modern. They adopt a pro-modern attitude toward progress, but not a modern assessment of anthropology because they reject dualism. With regard to posthumanism, the classification is even more problematic. Posthumanists are indifferent to the concept of progress, which is a decisive reason why they do not fit into Marquard's general scheme. In a certain sense, posthumanists also argue that it is better to live in a liberal-democratic society than in a medieval religious community. In this case, would it be appropriate to talk about political development and also of progress? If this were true, it would be true that posthumanists defend a certain kind of progress, but this argument would certainly not apply to other areas. For instance, they do not see any progress toward greater ultimate validity in the findings of the current natural sciences compared to the judgments of medieval Christians. With regard to ultimate truths, it is unclear from a posthumanist perspective whether the natural sciences, whose insights are accepted by many contemporary people, or Christianity, which was the prevailing worldview in Europe one thousand years ago, provide us with their insights. In any case, the question of a clear assessment of progress from a posthumanist perspective is difficult. Both affirmative and negative elements can be found in their assessment. Posthumanists defend a nondualistic ontology, but they cannot rule out the possibility that a Catholic dualistic ontology represents the truth in correspondence with reality, although they consider such an assessment implausible. In any case, Marquard's classification schemes of historical philosophies outlined above represent a perspective too narrowly constricted for today's ways of thinking. Nevertheless, they served a key function in my attempt to map out the uses and pedigrees of the various movements that strive to transcend humanism. The questions concerning the various topics beyond humanism are so complex and difficult, however, that many creative and intelligent thinkers are needed to develop and deal with appropriate approaches. The present introductory reflections on this nexus of topics represent only a few initial suggestions to promote the various intellectual reflections on the discourse beyond humanism.

CHAPTER 4

Nietzsche and Transhumanism

When examining transhumanism in depth, the similarity between transhumanist principles and those of Nietzsche's philosophy immediately becomes apparent. If you compare the concept of the posthuman and Nietzsche's overhuman (*Übermensch*), the similarities are especially striking. Transhumanist Nick Bostrom, however, rejects Nietzsche as a progenitor of transhumanism. He had recognized only some "surface-level similarities with the Nietzschean vision" (Bostrom 2005b, 4).

This rejection is probably based on the insight that it is usually unfavorable to be associated with Nietzsche, which is why Bostrom also insists that there are serious differences between transhumanism and Nietzsche's thinking. Unfortunately, Nietzsche is still regarded by many philosophers today as a mastermind of the Third Reich. This assessment of Nietzsche is incorrect, but the fact that he is perceived this way motivates many people not to avoid being associated with him, which is Bostrom's explicit rationale for rejecting Nietzsche.

In this chapter, I will show that Bostrom's assessment of Nietzsche's thought is inaccurate and that decisive similarities exist on a fundamental level between the posthuman and Nietzsche's overhuman. Habermas also shares this opinion, as he pointed out the similarities between these two worldviews. However, he considers both to be absurd. At least he describes transhumanists (whom he inaccurately calls posthumanists) as a group of crazy intellectuals who fortunately have not yet managed to get support from the general public for their elitist ideas.[1]

Nietzsche has also explained the significance of the supernatural with the help of a depth dimension that seems to be missing in

58

transhumanism but that is precisely what grants Nietzsche's position its enormous significance. To present my assessment, I proceed as follows. First, I compare the concept of the posthuman with that of Nietzsche's overhuman. I focus more on their similarities than on their differences. I also explain the context in which the overhuman attains its importance in Nietzsche, so that I can clarify the aspects that seem to me to be missing in transhumanism.

THE POSTHUMAN AND NIETZSCHE'S OVERHUMAN

Before I make a direct comparison between the posthuman and Nietzsche's overhuman (*Übermensch*), I will outline some fundamental principles of Bostrom's concept of transhumanism that are directly relevant to the concept of the posthuman. I compare these with corresponding principles within Nietzsche's philosophy. I am particularly interested in the evolution of the human being and of values, the natural sciences, self-enhancements, education, and the Renaissance genius.

Evolution of the Human Being

Both transhumanists and Nietzsche believe that nature and values are subject to constant change. Bostrom says the following on the topic of human nature: "Transhumanists view human nature as a work-in-progress." Nietzsche agrees with this view (Bostrom 2005c, 1). He develops a dynamic will-to-power ontology that applies to humans and all other beings and implies that all things are subject to constant change.[2] According to Nietzsche, all humans are organisms assembled from quanta of power, that is, of will-to-power constellations, structures that bear some similarities with Leibniz's monads (Sorgner 2007, 50). Like a monad, a quantum of power is a single entity that represents the basic element of the world. Unlike a monad, a quantum can interact with other quanta of power; it can grow and decay, it can feed itself (in metaphorical terms) and has a perspective on the world. On the basis of its perspective, the quantum decides how it acts, which in turn depends on its own possibilities and respective conception of power. Nietzsche uses a very broad and open-ended concept of "power." Any situation in which one quantum is stronger and more capable than another and in which it has the ability to dominate others is a form of power.

According to Nietzsche, all entities are composed of such constellations of power. The interaction of the various power relationships is also responsible for the evolution process that underlies the emergence of humans, animals, and plants. All kinds of organisms arose solely from the fact that the quest for power by the type of organism from which they originated was best fulfilled by a further evolutionary development. This process continued until humans were finally born, which does not mean that the evolution process was then complete.

The human species, like any other species, is of course not unchangeable. It developed and could pass away or develop in new ways. However, individual members of a species have a limited potential owing to their species affiliation. A species is characterized not only by the fact that its members can potentially reproduce with each other, but also by the fact that the members can develop only within the limits set by the species (Nietzsche 1988, 13:316–17).

Human beings therefore have only a limited number of possibilities for development since they belong to the human species and can develop only within the limited possibilities available within the species. As a rule, a person cannot pass on acquired qualities to his descendants. Under certain circumstances, however, Nietzsche advances such a "Lamarckism."

> One cannot erase from the soul of a human being what his ancestors liked most to do and did most constantly: whether they were, for example, assiduous savers and appurtenances of a desk and cash box, modest and bourgeois in their desires, modest also in their virtues; or whether they lived accustomed to commanding from dawn to dusk, fond of rough amusements and also perhaps of even rougher duties and responsibilities; or whether, finally, at some point they sacrificed ancient prerogatives of birth and possessions in order to live entirely for their faith—their "god"—as men of an inexorable and delicate conscience which blushes at every compromise. It is simply not possible that a human being should not have the qualities

and preferences of his parents and ancestors in his body, whatever appearances may suggest to the contrary. (Nietzsche 1989, 213–14)

This has practical implications. Nietzsche makes it clear, for example, that if a man enjoys nice meals and spending time in the company of women, then it is not advisable for his son to lead a chaste and ascetic life (Nietzsche 1988, 4:356–68).

Under special social and individual conditions, which Nietzsche does not describe exactly, evolution can take place, a process that is also discussed by transhumanists. Bostrom puts this in clear words: "A common understanding is that it would be naïve to think that the human condition and human nature will remain pretty much the same for very much longer" (Bostrom 2001). Nietzsche is not quite as optimistic as Bostrom, who posits that an evolutionary development of humanity will soon take place. However, he is on the same page with transhumanists in holding that such a development will happen if humanity does not die out.

Evolution of Values

Besides the ontological dynamism that can be found both in transhumanism and in Nietzsche's philosophy, they also both exhibit similar flexibility with regard to values. On this point Bostrom explains: "Transhumanism is a dynamic philosophy, intended to evolve as new information becomes available or challenges emerge. One transhumanist value is therefore to cultivate a questioning attitude and a willingness to revise one's beliefs and assumptions" (Bostrom 2001).

Nietzsche makes a similar point to Bostrom's, that prevailing values have constantly changed. He describes his interpretation of the evolution of values in *The Genealogy of Morals* (1887). Values are apt to change on cultural, social, and personal levels. The power concept bound to a quantum of power could also change on the basis of new experiences and insights. The content of the concepts of power is always linked to the perspective of power wielders (Sorgner 2007, 79–83). There are no absolute and unchanging values since he rejects the existence of the

Platonic world of ideas that would be the necessary foundation for any such permanent values.

Natural Sciences, Enhancements, and Education

Nietzsche and the transhumanists both have a worldview that differs significantly from the traditional Christian worldview and from any worldview that inherits Christian values. Since even today the thinking of many people in the Western world is still shaped by Christian values, Nietzsche and the transhumanists are trying to reevaluate the values.

Bostrom stresses: "Transhumanists insist that our received moral precepts and intuitions are not in general sufficient to guide policy" (Bostrom 2001). For this reason, he proposes values that take a dynamic worldview into consideration: "We can thus include in our list of transhumanist values that of promoting understanding of where we are and where we are headed. This value encloses others: critical thinking, open-mindedness, scientific inquiry, and open discussion are all important helps for increasing society's intellectual readiness (Bostrom 2001). Nietzsche largely concurs with this statement. He had great respect for critical thinking—many consider him to be one of the sharpest critics of morality and religion. He also respects scientific research (Sorgner 2007, 140–45). In numerous passages he emphasizes that the future will be shaped by the scientific spirit, which is also the reason why his philosophy has fared so well. His way of thinking appeals to scientifically minded people.

Nietzsche's esteem for the natural sciences is recognized by many Nietzsche scholars (Moore and Brobjer 2004; Birx 2006). His doctrine of the eternal return is based on premises introduced by natural scientists (Sorgner 2007, 65–76). His will-to-power anthropology agrees in numerous aspects with scientific theories of the human. Although he is often critical of Darwin's theory of evolution, he himself espouses a similar one. This is because of the fact that he is usually particularly critical of precisely those thinkers who are most akin to him (Nietzsche 1988, 8:97). Nietzsche criticizes Darwin specifically for positing the struggle for survival. Nietzsche, by contrast, argues that the striving for power alone is the basis of interpersonal struggle (Sorgner 2007, 62). People strive for power. The struggle for survival is only one marginal expression of the struggle for power, according to Nietzsche.

Enhancement is in the interest of everyone who strives for power. Transhumanists also argue for enhancement. Transhumanists advocate technologies and other tools that can be used for the "enhancement of human intellectual, physical, and emotional capacities" (Bostrom 2001) so that the emergence of posthumans can be prepared. For this reason, transhumanists also appreciate a kind of liberalism that implies that people have the right "to live much longer and healthier lives, to enhance their memory and other intellectual faculties, to refine their emotional experiences and subjective sense of well-being, and generally to achieve a greater degree of control over their own lives" (Bostrom 2005c, 1).

Bostrom has given a detailed picture of exactly what these demands imply. It is important to stress that Bostrom, unlike Habermas, argues that parents have the right to dispose of the genetic composition of their own children. Because of the great importance of Habermas's position, I present in a brief excursus the transhumanist critique of his position.

Habermas distinguishes between children who have become and children who have been made (Habermas 2001, 41, 45, 80–93) and takes the view that, first, it is immoral for parents to determine the genetic composition of children, since children are thus forced in one direction, which is not the case if they are the result of accidental processes (Habermas 2001, 53–55). Second, there is a significant difference between upbringing and genetic enhancement (Habermas 2001, 31, 87–114). In the context of education, children have the opportunity to do something to oppose the results of their upbringing (Habermas 2001, 100). Besides that, one can still change the characteristics brought about by education, which in the case of genetic composition should not be the case (Habermas 2001, 111). These two factors are used to treat education and genetic enhancement as morally different.

Bostrom, on the other hand, stresses the following: "Transhumanists also hold that there is no special ethical merit in playing genetic roulette. Letting chance determine the genetic identity of our children may spare us from directly confronting some difficult choices" (Bostrom 2001).

This means that he rejects Habermas's first point. Bostrom implicitly suggests that most parents love their children and work hard to ensure that good things come to their children. What they consider good for their children may be something they embrace wholeheartedly or just something that is in their child's interest. Irrespective of the parents' concept of good, the decision that the parents make for their child might generally be better than the result based on genetic roulette or chance. Then even Habermas's second argument would not be tenable. If the parents choose a better genetic design than the one chance would have determined, then it is not morally condemnable if that design is harder to change than the traits garnered from education. Perhaps we should even go as far as saying that children are better off when traits in their interest are irreversible. That said, it is far from clear whether Habermas's irreversibility thesis is even correct. There is solid scientific evidence that the qualities acquired through upbringing can be so firmly integrated into one's own personality that they too cannot be decisively changed. A further possibility is that the genetically modified traits are just as reversible as the traits acquired through education. Both variant hypotheses are quite plausible and are supported by the latest scientific research (Sorgner 2015).

Critics of enhancement further stress the dangers associated with new techniques. They cry out that some modifications are certain to fail initially. It is reprehensible to play around with people and to use them exclusively as means.[3] Bostrom discusses the consequences of the transhumanist rejection of fear: "Transhumanism tends toward pragmatism . . . taking a constructive, problem-solving approach to challenges, favouring methods that experience tells us give good results, and taking the initiative to 'do something about it' rather than just sit around complaining" (Bostrom 2001).

All natural scientists and explorers who want to explore new fields must be courageous because they are venturing into potentially dangerous regions. The same applies to researchers in the field of genetic engineering. America would probably never have been discovered if there hadn't been brave people who had taken the risks to carry out the discoveries. There are therefore good reasons for giving more weight to the principle of individual initiative, or Max More's proactionary principle, than to the precautionary principle.

Courage is a central virtue among the moral concepts Nietzsche upholds. Nietzsche also emphasizes that the natural sciences will be of central relevance in future centuries. This development is in the interest of his philosophical program. On these grounds, it cannot be ruled out that Nietzsche would have been in favor of implementing genetic engineering measures, even if he emphasizes the role of education in the evolutionary development of the overhuman. If genetic engineering measures or liberal eugenics can be seen as a special kind of education, which many transhumanists advocate, then it is possible that genetic engineering would also have been embraced by Nietzsche since education plays an important role in his ethics, and since he champions the natural sciences, self-overcoming processes, and the emergence of the overhuman.

A Perspectival Theory of Values and the Renaissance Genius
Transhumanists do not intend to impose their values on other people since "transhumanists place emphasis on individual freedom and individual choice in the area of enhancement technologies" (Bostrom 2005c). One reason for this premise, according to Bostrom, is the following: it is "a fact that humans differ widely in their conceptions of what their own perfection would consist in" (Bostrom 2001). And furthermore: "The second reason for this element of individualism is the poor track record of collective decision-making in the domain of human enhancement. The eugenics movement, for example, is thoroughly discredited."[4] What is decisive here is that transhumanism is associated with a perspectival theory of values, similar to the one Nietzsche upholds.

According to Nietzsche, every constellation of power, and therefore all human beings, have their own perspective on the world, their own hierarchies of values, and their own concepts of power. Since every conception of power varies depending on who one is and what one's life story has been, every human being has a unique, and thus a special, idea of his own concept of "perfection." This statement also applies to Nietzsche himself. His own ideal of power is closely connected with the feeling of power tied to the classical ideal (Sorgner 2007, 53–58.). A similar ideal is represented by transhumanists, about whom Bostrom says the following: "Transhumanism imports from secular humanism the ideal of the fully developed and well-rounded personality. We can't all be renaissance geniuses, but we can strive to constantly refine ourselves and to broaden our intellectual horizons" (Bostrom 2001). Nietzsche not only develops the notion of the "fully developed and well-rounded personality," but he also demands that we "constantly refine ourselves and expand our intellectual horizons." Nietzsche describes this process as self-overcoming (Nietzsche 1988, 4:146–49). Higher human beings constantly want to overcome themselves in order to become stronger with respect to the different qualities to be enhanced in human beings, so that ultimately the overhuman can emerge. In transhumanist thinking, the overhuman is called the posthuman.

The Posthuman, the Transhuman, and Nietzsche's Overhuman
Who is the posthuman? What qualities does he, she, they, or it have? I think that many transhumanists ultimately agree with the following statement: "We lack the capacity to form a realistic intuitive understanding of what it would be like to be posthuman" (Bostrom 2001). However, various transhumanists have tried to describe the posthuman more precisely. In order to present the range of transhumanist theories of the posthuman, I present both Bostrom's and Esfandiary's models.

According to Bostrom, F. M. Esfandiary's model is as follows: "A transhuman is a 'transitional human,' someone who by virtue of their technology usage, cultural values, and lifestyle constitutes an evolutionary link to the coming era of posthumanity" (Esfandiary in Bostrom 2005b, 12). This position implies that the transhuman still belongs to the human species but has already attained qualities that go beyond the usual concept of the human and has the potential to initiate the evolutionary step toward a new species. The new species is referred to here as the posthuman. This conception implies that humans and transhumans can still reproduce together, but that this is not the case with posthumans, just as we cannot reproduce with apes. It goes without saying that posthumans would depend on technology for reproduction.

Bostrom uses the term "transhumanism" to refer to dynamic philosophies that approve of new developments and scientific research. His

66

concept of the posthuman is closely linked to the following consideration: "By a posthuman capacity, I mean a general central capacity greatly exceeding the maximum attainable by any current human being without recourse to new technological means" (Bostrom 2009, 108). Posthuman qualities are thus not to be identified with those that present humans possess. He claims, however, that we can transform ourselves into such beings.[5] He does not mean to say that we will attain qualities that cannot be compared with those of the presently living human beings, but rather that each individual human being can become a posthuman with the support of technological aid. He even asserts the following: "This could make it possible for personal identity to be preserved during the transformation into posthuman" (Bostrom 2009, 123). According to him, both humans and posthumans belong to the human species, which implies that they would have the capacity to bear offspring together without any technical assistance. Posthumans are therefore not members of a separate species, but a special form of the species "human," although this potential form of the species would possess properties that we cannot even imagine. We mere humans would also have the potential to achieve these qualities through emotional, physiological, and intellectual enhancement. Moreover, Bostrom considers it unlikely that such enhancement would be possible without technological assistance.[6]

At this point I would like to lay out some of the general possibilities for enhancement that Bostrom endorses. Here the term "eugenics" gains renewed relevance, even though Bostrom rejects it. By a minimal definition, "eugenics" means nothing other than enhancement of hereditary predispositions, a program many transhumanists advocate (Sorgner 2006b, 202). Bostrom's rejection of the word "eugenics" is certainly connected with the fact that it is always associated with the abominable procedures implemented under the Third Reich. However, the eugenics practiced at that time was state-regulated. The state-regulated eugenics of the Third Reich is morally reprehensible, and no one today seriously defends that policy within the academic bioethics debate. Liberal eugenics, on the other hand, has been intensively discussed since the beginning of this millennium, and ethicists, such as Nicholas Agar (1998), support the implementation of numerous methods that belong to liberal eugenics. Transhumanists, too, regard these methods as morally legitimate forms of human enhancement. Both state-regulated and liberal eugenics encompass heteronomous forms of eugenics, which implies that people decide on the enhancement of other people. In state-regulated eugenics, the state decides, whereas in liberal eugenics parents have the right to decide what to do with their offspring. Many transhumanists particularly emphasize the relevance of autonomous genetic modification, which is also a variant of liberal eugenics (Sorgner 2006b, 205). It implies that people have the right to morphological freedom, including the right to decide for themselves whether or not they want to be modified by genetic engineering measures. In order to avoid false associations, many bioliberals and transhumanists today speak of genetic enhancement rather than liberal eugenics.

Through technologies for self-fashioning, the day could come when humans transform into posthumans. According to Bostrom, this should in fact be the preferred method for bringing about posthumans:

> It follows trivially from the definition of "posthuman" given in this paper that we are not posthuman at the time of writing. It does not follow, at least not in any obvious way, that a posthuman could not also remain a human being. Whether or not this is so depends on what meaning we assign to the word "human." One might well take an expansive view of what it means to be human, in which case "posthuman" is to be understood as denoting a certain possible type of human mode of being—if I am right, an exceedingly worthwhile type. (Bostrom 2009, 135)

In light of the above concepts of the posthuman, I conclude that Nietzsche's concepts of higher humanity and of the overhuman resemble Esfandiary's concepts of the trans- and posthuman, but not Bostrom's. According to Nietzsche, members of the human species can usually develop characteristics only within the limits of our species. Under certain conditions, however, evolutionary processes can occur. Again, according to Nietzsche, evolution does not progress continuously, but in leaps and bounds. If the conditions within a species are favorable for an evolutionary leap, then members of a new species could be born to multiple parents at the same time. The couples from whom the over-human is born must possess characteristics of higher humanity. Under normal circumstances, a higher human could not pass down higher traits, but if many higher humans appear at the same time, and other favorable conditions are met, then such an evolutionary leap could occur (Nietzsche 1988, 13:316–17).

According to Nietzsche, higher humans still belong to the human species, but they also possess qualities that overhumans could possess. In contrast to overhumans, higher humans cannot pass on their special qualities through reproduction. A higher human being would arise only by chance. The necessary prerequisites for their emergence (elders, cultural environment, and education) must be in place. One of Nietzsche's prime examples of a higher human is Johann Wolfgang von Goethe (Nietzsche 1988, 6:151–52). Higher humans must have had a special nature that enabled them to become such beings. Attaining this special nature, however, presupposes that they work hard to enhance themselves, especially to enhance their intellectual abilities, and here Nietzsche puts a special emphasis on their skill in interpretation. Nietzsche does not refer to technological possibilities for enhancement—Bostrom is right about that (Bostrom 2005b, 4)—but Nietzsche does not exclude such methods either.

A decisive difference between overhumans and higher humans lies in their respective potentials. Potential in both cases corresponds to that of the given species. Up until now there has not been an overhuman, but the qualities of a normal overhuman lie beyond those of a higher human being. The overhuman arises by means of an evolutionary step, which propels them out from the group of the higher humans. Nietzsche does not exclude the use of technology to elicit this event. Although his statements are rather vague, his basic attitude is similar to that of the transhumanists. In any case, he predicts that the future time will be ruled by the scientific spirit, which will bring an end to the dominance of dualistic worldviews and the increased adoption of the basic principles of the natural sciences.

After having briefly described Nietzsche's concepts, it is worth mentioning how closely Nietzsche's higher humans resemble Esfandiary's concept of the transhuman, and that Nietzsche's overhuman resembles

Esfandiary's concept of the posthuman in many respects. What about Bostrom's concepts of the trans- and posthuman?

Bostrom says the following on the topic: "One might well take an expansive view of what it means to be human, in which case 'posthuman' is to be understood as denoting a certain possible type of human mode of being" (Bostrom 2009, 135). He thus implies that posthuman abilities cannot be found in currently living people. Since Nietzsche argues that each species is determined by its clear boundaries, it is implausible to claim that his philosophy can be reconciled with the stance that humans may one day possess properties that cannot even be imagined today. For this reason, we should gladly conclude that Nietzsche's concept of the overhuman cannot be reconciled with all transhumanist concepts of the posthuman, although he shares many basic anthropological and ethical attitudes with transhumanists. However, there are transhumanists whose concept of the posthuman is very similar to Nietzsche's overhuman.

THE OVERHUMAN AND NIETZSCHE'S HOPE FOR THE FUTURE

Transhumanists have so far not argued for a particular meaning to life, for why they represent the values they uphold, and for why the posthuman should emerge. The closest they come to these topics is to encourage an overall enhancement in quality of life. Nietzsche explains the significance of the overhuman within his philosophy. This can even be seen as the fundamental rationale behind his thinking.

According to Nietzsche, philosophers are creators of values that are ultimately based on their personal prejudices.[7] He sees his own prejudices as corresponding to the demands of the spirit that will dominate the coming centuries of Western Europe. Here "spirit" does not refer to an immaterial *nous* in the Platonic sense or to any haunting ghosts, but rather to a part of the body in which the capacity for interpretation is seated, the characteristics of which in turn are based on the physiological strength of the body. He distinguishes between a scientific and a religious spirit. Weak, reactive people who cannot fulfill their desires in this world are endowed with a religious spirit that makes them long for the good life in the hereafter. This spirit had long been the predominant one. Slowly, however, people had become stronger, and the scientific spirit had developed in them. The importance of this spirit had grown enormously, especially since the Renaissance. Nietzsche expects that this spirit will become even more important in the future. Since he regards his worldview as plausible within the scientific spirit, he expects its plausibility to increase as time goes on.

Nietzsche viewed Plato as the representative of a philosophy based on the religious spirit. His own way of thinking, on the other hand, was based on the scientific spirit. He described Christianity, which had determined thought and action in the Western world for a long time, as Platonism for the people. His own philosophy is an inverted Platonism. Just as the Christian worldview had dominated numerous centuries, so his scientific worldview should govern future centuries. This is why Nietzsche repositions the central building blocks of Christian-Platonic thought in reverse.

A central aspect of Christian teaching, according to Nietzsche, is that of personal survival after death. This appeals to many people and gives their lives meaning. If my brief summary is correct, then Nietzsche must also have a reversed understanding of the afterlife, which gives meaning back to life before death. At this point the overhuman becomes relevant, along with the doctrine of the eternal return. I will not be able to argue here about his doctrine of eternal return, his "theory of salvation," although it is closely connected with the concept of the overhuman.

For Nietzsche, the overhuman is the meaning of the earth. The overhuman thus represents the meaning-giving concept within his worldview, which is to replace the Christian idea of the afterlife. It is in the interest of higher men always to overcome themselves. The highest form of overcoming is thus to be seen in overcoming the human species, and everyone who has been constantly occupied with overcoming themselves can regard themselves as ancestors of the overhuman. In this way, the overhuman exists to give meaning to human beings. It is, however, not a transcendent but rather an immanent sense that is suitable for people who think in a natural-scientific way and who have given up their faith in a world beyond. Carl Jung stresses that "man cannot stand a meaningless life" (Jung in Stevens 1994, 126). Nietzsche and Plato would agree with this statement. Such a dimension seems to me to be included within the concept of the posthuman. In another context Bostrom explains the following: "Many people who hold religious beliefs are already accustomed to the prospect of an extremely radical transformation into a kind of posthuman being, which is expected to take place after the termination of their current physical incarnation. Most of those who hold such a view also hold that the transformation could be very good for the person who is transformed" (Bostrom 2009, 126).

I suspect that the transhumanist concept of the posthuman cannot be fully understood without considering its meaning-giving character: "I want to be the ancestor of the posthuman." And yet I doubt that many transhumanists would agree with this statement. They would probably be worried that it was mixing scientific and religious categories. It is crucial to recognize, however, that unlike the Christian concept of the afterlife, what I take for the most meaningful aspect of the posthuman is its basis in scientific hopes, the appreciation of the sensory world, and an immanent goal; recognizing this basis can make the concept of the posthuman even stronger. Nietzsche regarded the overhuman as the meaning of the earth. From my point of view the meaning of the posthuman can be understood only if one acknowledges that it is a meaningful concept, which gives meaning to life for scientifically oriented people.

I myself share the opinion held by Nietzsche and other transhumanists that human beings will develop into posthumans, that people are part of the evolutionary chain. I also believe that important human modifications can always be realized through technological developments. However, I am rather skeptical about the assessment that radical developments of the human constitution are necessarily imminent. I consider it unrealistic that technological singularity, that is, the decoupling of the development processes of artificial intelligence from human influences, will take place in thirty years. In such forecasts, even so-called mind uploading is supposedly almost achievable. I do not share this assessment. Rather, I expect decisive developments to occur in the field of genetic engineering. I value the biotechnological possibilities that could enhance human beings in such a way that in the foreseeable future there will be developments that will seriously alter human capacities. Here one does not have to be a prophet, if one is of the opinion that with each change in human ways of life, new dangers will arise at the same time. The fundamental human questions, which concern matters of ultimate concern (death, judgment, God, immortality), will not be answered with the help of biotechnological developments and will also not disappear. However, I assume that numerous technological developments will result in numerous enhancements in the ways we live life, such as curing illnesses.

Transhumanists tout technological processes of self-enhancement, seeing as human beings have constantly developed such innovations anew in the course of their phylogeny, with one difference: transhumanists rally behind these processes with the specific aim of increasing the probability of the emergence of the posthuman. A strong form of transhumanism contends that there is a moral, but no legal, obligation to enhance (e.g., Savulescu). On the other hand, a weak form of transhumanism defends only that enhancement techniques increase the probability of a good life for many people. I share the latter assessment.

CHAPTER 5

Twelve Pillars of Transhumanist Discourse

I stumbled on transhumanism through my own work on Nietzsche (Sorgner 2009). Even though some transhumanists certainly deny Nietzsche's forefather status (Bostrom 2005a, 4), I am not alone in seeing the philosophical relationship between the two schools of thought. Max More's transhumanism was also decisively influenced by Nietzsche's philosophy (More 2010), an influence that has continued since he and FM-2030 worked out the fundamental principles of transhumanism across its contemporary variants over thirty-five years ago (Ranisch and Sorgner 2014).

The term "transhumanism" was, as already mentioned, coined by Julian Huxley in the 1951 essay "Knowledge, Morality, and Destiny" (Huxley 1957). Julian Huxley was not only the brother of Aldous Huxley (*Brave New World*), an author who was himself critical of technology. He was also the grandson of "Darwin's Bulldog" Thomas Henry Huxley, half-brother of Nobel Prize winner Andrew Fielding Huxley, long-standing chairman of the "British Eugenics Society," and the founding director of UNESCO. This cursory review of the founding figures of transhumanism reveals the movement's intellectual affinity with English philosophy, naturalism, utilitarianism, and evolutionary theory. One book by Julian Huxley even bears the title *Evolutionary Humanism*, which explains why his thinking is also referred to as secular humanism, a characterization that is also certainly correct. But is Julian Huxley a humanist or a transhumanist thinker? What is the relationship between humanism and transhumanism? The answers depend on how we understand the word "humanism." Do secular, naturalistic, and evolutionary humanisms actually deserve the name? I do not think so, because these pseudohumanisms do not carry the dualistic implications of the metaphysical concept of "humanism" discussed above.

Transhumanism is often referred to as hyperhumanism. I find this description inadequate. That assessment derives in part from a misunderstanding about the transhumanist motif of mind uploading. Those who call transhumanism hyperhumanism imagine that the process of mind uploading presupposes a dualistic theory of the human. However, this assessment is incorrect. Transhumanists usually purport a naturalistic anthropology. Mind uploading presupposes no dualistic anthropology since it can also be advocated plausibly in the context of naturalism. However, this can be so only if we accept the validity of a functional theory of mind (More 2013, 7). And the validity of that theory remains a legitimate question. What is decisive in this context, however, is that since transhumanists advance a naturalistic anthropology, it becomes highly questionable to identify transhumanism as hyperhumanism. Before I present the twelve pillars of my own weak Nietzschean transhumanism, some further traits of this basic philosophical attitude must be illustrated.

The decisive core characteristic of transhumanism is the approval of the use of technologies to increase the probability that the posthuman will emerge. Although there is no substantive description of the posthuman shared by all transhumanists, posthumans are always understood to be a further development on humanity. The most prominent variants for this are the hope for a silicon-based future, which could be realized by mind uploading, or the goal of further carbon-based development, where it remains open whether posthumans will still be members of the human species or whether they will possess capabilities that fundamentally exceed the capabilities currently possessed by living humans. People who find themselves in the process of this development toward the posthuman are called transhumans. This term was introduced by FM-2030 as early as 1974 in an article speculating about the living conditions that women would enjoy in the year 2000 (Esfandiary 1974). In any case, transhumanists argue that better developed abilities will increase the probability of leading a good life.

Who posthumans will actually be remains an open question. Their emergence is probable within a naturalistic model of human life since everything that has arisen will either continue to develop or die out. In order to avoid extinction, it is necessary to constantly adapt in new ways to changing environmental conditions. These considerations also include the possibility that posthumans could be personalities uploaded onto a computer.

I cannot rule out the possibility of mind uploading. Director Wally Pfister presents it in a vivid and impressive way in the 2014 film *Transcendence*. At the same time, further development in carbon-based existence seems much more likely to me to be realized within a manageable timeframe, so long as humanity does not die out beforehand. Transhumanists take the question of existential risks to humanity very seriously and discuss them intensively on the basis of solid scientific research, especially since numerous scientists at some of the world's leading universities call themselves transhumanists, and some have devoted a great deal of attention to this question. The possibility of the extinction of humanity is also placed at the center of the action in the first popular novel to respond directly to transhumanism (*Inferno* by Dan Brown). This novel presents transhumanism in a fair way and points to some of the central challenges facing us today.[1]

Through the following twelve pillars of transhumanist discourse, I will present foundational transhumanist positions and wide-ranging reflections on particular questions. In this way, the widely misappropriated heterogeneity of transhumanist thinking becomes clear. In addition to addressing hotly debated questions, this section also enumerates the twelve pillars of my own weak Nietzschean transhumanism.[2]

1. REJECTION OF THE CATEGORICAL EXCEPTIONALISM OF THE HUMAN

Like a transhumanist, Nietzsche understands the human as a constantly self-overcoming being that constitutes a bridge between animal and overhuman, a correspondence that allows us to ascertain some parallels between Nietzsche's overhuman and the transhumanists' vision of the posthuman. There are also parallels between Nietzsche's higher humans and the transhumanists' transhumans (IEET). The backbone of these Nietzschean, transhumanist reflections is a concept of humanity that differs categorically from the one that still prevails in most contemporary liberal legal systems. I have drawn attention to this in my monograph "Human dignity after/according to Nietzsche" (Sorgner 2010b). In October 2014, a collection was published in which theologians, ethicists, philosophers, anthropologists, jurists, and religious scientists responded to my own critical arguments. The decisive question here is the following: Can a metaphysical anthropology be the foundation for a pluralistic, liberal-democratic state like Germany's? I for one doubt it (Sorgner 2014b).

In the two publications just mentioned, I make it clear that a categorical dualist human image prevails on a legal level in Germany as well as in many other countries because of our Christian and Kantian heritage. Animals are no longer regarded as mere things, but they are still legally treated as things. The characteristic of dignity belongs exclusively to human beings. Fortunately, this line of thought, which is prevalent in legal reasoning, is one that many people no longer share and that, for this among other reasons, should no longer be binding within the framework of the constitution. Herewith I plead for a new modesty about what it means to be human, according to which human beings no longer see themselves as the singular crown of creation but recognize that humans differ only gradually from other living beings. On the legal level, it may be more appropriate to avoid ontological fixations based on a theory of humanity (*anthropologische Theorie*) and instead to refer more directly to shared norms and values.

On this point, it may be helpful to note that rejecting human beings' special categorical position does not imply that one must necessarily reject the possibility that any particular species can claim a unique status owing to special characteristics. Humans might seem to have a special status because of our ability to learn one or more human languages. But if that were so, then South American vampire bats could also have a special status as the only mammals able to feed on blood alone, and who may even be able to detect blood by infrared sensors located in their noses. Decisive for the argument presented here is the dissolution of the *categorical* ontological special status of humans. The result of this dissolution is the assertion that natural living beings differ from each other only by degrees, an assertion accompanied, for example, by the assessment that it is implausible that humans consist of an immaterial soul and a material body. How to take such an assessment into account in a liberal-democratic state, however, is a complex question of policy and law. I consider the legal relevance of the belief in the soul indisputable since it remains such a widely held belief. How to take it into account in Germany depends on how relevant we decide religious beliefs are when interpreting the German constitution. The fact that religious perspectives are given special consideration within the framework of the Basic Law for the Federal Republic of Germany becomes clear on the basis of the following words, which can be found at the beginning of the preamble to the Basic Law: "Conscious of their responsibility before God and man." The legal relevance of these sentences in practice is a separate question. There is the possibility that they refer to the special position of Christianity. It could also be that they imply that all religious worldviews should be given special consideration proportional to their social anchoring, that is, according to the extent to which they resonate within a society. That naturalistic or evolutionary thinking represents an equally legitimate worldview within society seems questionable. Whether the current preamble is even in harmony with liberal, pluralistic understandings of democracy is open to doubt. Most probably, only a thoroughly secular society could guarantee the appropriate conditions for liberalism, pluralism, and democracy. In order to establish such a view legally in Germany, larger revisions would have to be carried out.

The arguments briefly presented here are particularly relevant for Germany—whose constitution has an unusually strong human rights focus—for historical reasons. In the international context, transhumanists primarily argue for *nonhuman personhood*. In this context, they are close to numerous interpretations developed by Peter Singer.

2. REJECTION OF THE PROHIBITION AGAINST INSTRUMENTALIZATION

At the same time, to revise the currently dominant notion of humanity also means that another Kantian legacy should be revised. The prohibition against instrumentalizing another human being that Kant imposed still has important practical implications. When ruling for the prohibition against the shooting down of hijacked aircraft or the prohibition against peep shows, explicit reference was made to instrumentalization. Hijacked aircraft may not be shot down in Germany as long as innocent people are on board since the instrumentalization of innocent people is deemed immoral because of their human dignity. This is the case even if there is a threat that the aircraft will head for a nuclear power plant where a crash would result in the death of millions of people in addition to the innocent people on board. What is interesting for the reflections addressed in this context is that the dualistic categorical-substance differentiation of things from persons, that is, the differentiation of human beings who have dignity from mere objects, is a prerequisite for the application of this prohibition. The prohibition requires that people not be used exclusively as means to an end. Only things may be used exclusively as means. However, if the categorical duality of substance between persons and things is to be dissolved, or if ontological assessments should be excluded from consideration in legislation, then the ban on instrumentalization would also have to be revised. This is not to say that on a legal level the duality of persons and things should be replaced with a gradual distinction. Such a change would only replace one anthropology with another. I do not see how the legal validity of a strong image of the human can be reconciled with the founding principles of the liberal, democratic, and pluralistic state. It seems more appropriate to me to ban strong ontological statements from the law and to replace them with discursive norms. Prohibiting instrumentalization also plays a central role in the debates on preimplantation diagnostics, PGD for short (Sorgner 2013c).

The present argument against the instrumentalization prohibition was developed by me and is of particular relevance for German-language discourses. In the English-speaking ethical context, utilitarianism plays a much more important role than Kant's ethics in justifying the prohibition of instrumentalization, which has a legal context in both cases. These considerations do not suggest that utilitarian considerations should play a more important role in the German legal context in the future. Rather, they point to a challenge whose consequences can hardly be estimated.

It is problematic that a strong Christian and Kantian anthropology is the universally valid legal basis because of its paternalistic implications. If an anthropology based on a gradual distinction between humans and animals and also between humans and AI were to be enshrined in law, this could have the following implications for the prohibition on the instrumentalization of human lives. Either the total instrumentalization of human lives could gain moral legitimacy, or the total instrumentalization of grass, flowers, and strawberries could become morally illegitimate. Both implications would be implausible and would have problematic practical consequences. Moreover, to introduce restrictions on the instrumentalization of nonhuman animals would not represent the most thorough way of thinking through this transhumanist insight anyway since the prohibition against instrumentalization in itself presupposes the separate existence of persons and things. Under new anthropological conditions, this prohibition could no longer be applied meaningfully. For this reason and primarily because every strong anthropology introduces paternalistic consequences, which are incompatible with a liberal, democratic state, I plead for banishing strong ontological and anthropological judgments from the law in favor of contingent values and norms that can be revised as conditions change.

3. SELECTION AS A MORALLY LEGITIMATE OPTION

In the last few years in particular, preimplantation genetic diagnosis (PGD) has been intensively discussed in the media. Because of the very rigid German embryo protection law, PGD by blastomere biopsy is still forbidden because totipotent cells are examined and therefore destroyed. Only the blastocyst biopsy, in which only trophoblasts are examined, can be used under very special conditions, that is, when the child is at high risk of inheriting a severe hereditary illness. This legal regulation is highly questionable morally, especially when it is taken into account that an argument against PGD is based on the claim that this process promotes the stigmatization of people with disabilities. The stigmatization of people with disabilities seems to me, however, to be promoted more by the existing legal regulation than would be the case with a more liberal regulation of PGD. Finally, the currently valid regulation states that a legally prohibited process—the examination and subsequent selection of a fertilized egg cell—may be carried out if this can prevent the birth of a person with a severe disability. This legislation suggests that life with a disability is so terrible that even an otherwise applicable rule can be suspended to prevent it from occurring. However, it is not only the moral problem of stigmatization that speaks against this legislation. A philosophical consideration also suggests a different approach to the question.

If we were to evaluate structurally analogous procedures from a morally analogous perspective, which I would consider philosophically appropriate, then the selection of fertilized oocytes should be regarded as legally legitimate after both blastomere and blastocyst biopsy, since these processes can be regarded as analogous to the choice of partner for propagation purposes. In both cases we select a certain number of genetic possibilities of our offspring—after all, 50 percent of the genes of our own offspring originate from the father and the remaining 50 percent from the mother. Even in the case of partner selection for reproductive purposes, our offspring have their genetic potential constrained by our partner's genetic makeup. In both cases, however, it is not the case that one's own offspring is genetically determined, since genetic changes also occur through epigenetic processes during development. I consider it highly problematic to treat these two processes as morally different (Sorgner 2014a). The fact that the equal treatment of the two aforementioned processes can also prevent the stigmatization of people with disabilities underlies the solution I have proposed to this question.[3]

This concern is also of primary relevance for the German-language debate. The structural analogy argument I have developed illustrates how implausible the moral reasoning behind this legislation is in Germany. In the English-speaking debate this question is also handled much more bioliberally than is the case in German-speaking countries. It is a widespread basic attitude in English-speaking countries that selection is morally unproblematic. Julian Savulescu even argues that there is a moral obligation to select those fertilized oocytes that have the greatest probability of leading a good life within the framework of artificial

insemination and subsequent preimplantation diagnostics. I for one see
paternalistic implications in this moral demand (Sorgner 2014a).

4. GENETIC MODIFICATION AS FUTURE VARIATION ON UPBRINGING

A similar argument by analogy can be made for genetic modification, a
process that is considered morally reprehensible by all leading German
ethicists and philosophers, such as Sloterdijk and Habermas.[4] In my
view, this assessment is also untenable, since traditional education and
genetic enhancement by modification should be seen as structurally
analogous processes (Sorgner 2010a, section 1.1.1). The possibility of
treating these processes as parallel is first of all given by the fact that in
both cases parents are responsible for changes in their offspring, and
only in the context of this special relationship can serious consideration
be given to whether people should be allowed to make such important
decisions for the lives of others. Habermas opposed this consideration
by arguing that genetic changes must be unchangeable, whereas traits
produced by upbringing can always be reversible.

Both of Habermas's assessments are inaccurate. Scientific investigations have meanwhile made it clear that genetic as well as educational
changes can be both reversible and irreversible, which is why both processes are parental decisions that can influence one's own offspring in
different ways. The fact that genetic modifications can be successful
and do not necessarily entail unwanted side effects is illustrated by the
example of a successful gene therapy for a hereditary eye disease carried
out in 2007 at the hospital of University College London (Maguire et al.
2008). Since the distinction between therapy and enhancement is not
categorical, this example is significant for any type of genetic modification. Genetic predispositions are therefore not necessarily unchangeable
but can in some cases be changed in a reliable way without side effects.
The fact that posttraumatic stress disorder can become a permanent
personality disorder after being caused by the experience of violence,
for example, makes it clear that environmental changes can also cause
irreversible states (Rentrop, Müller, and Baeuml 2009, 373). Educational

influence is a special form of environmental influence. The example therefore illustrates that environmental influences can in principle also produce unchangeable states. Education is, by the way, probably one of the longest established and most widespread enhancement techniques. Vaccination is also a widely accepted enhancement technique whose significance can hardly be underestimated since vaccination produces a characteristic in the human body that was not previously present. As a rule, education is even obligatory as a technique for the enhancement of others. The legal obligation to obtain certain vaccinations was abolished in Germany at the beginning of the 1980s, but still exists in many other countries such as the United States and Italy. Perhaps compulsory schooling in Germany should also be converted into compulsory education (where homeschooling is an alternative to school attendance), as is the case in many other western countries, such as the United States.

Bioliberal and transhumanist authors have long argued that parent-determined genetic modifications are an extension of traditional educational measures (Robertson 1994, 167). The fact that this is the case has, however, only ever been postulated by a few thinkers. In an article published in 2015, I presented a detailed argument that clearly illustrates the structural analogy between the two processes (Sorgner 2015). Especially in light of the recently developed efficacious and cost-effective genome editing method CRISPR-Cas9, the importance of the question of morally evaluating genetic modifications has once again become directly relevant. This has also added the question of the assessment of germ line interventions, where the consequences of potential interventions would impact not only the next generation but further generations as well.

5. NEGATIVE FREEDOM AS CENTRAL ACHIEVEMENT

The demands raised here are based on the insight that negative freedom, that is, the absence of coercion (freedom from), is a desirable and affirmable achievement. Positive freedom, on the other hand, represents the freedom to do something, and with it the ability to make use of negative freedom. In order to be free to communicate openly, I need not only freedom of expression as negative freedom but also access to the necessary means of communication, that is, possibly certain financial means that enable me to own and use a smartphone and thereby communicate my opinion in an appropriate way. The primary political norm addressed here is that of negative freedom. It does not represent an ultimate truth, but that does not detract from its historical rarity. It was won during the Enlightenment within the framework of struggles waged on numerous levels. People with a variety of social allegiances and disciplines fought during the Enlightenment to limit state and church oppression. For a long time, religious and political rulers were entitled to decide that conception of the good their subordinates would have to live by, and even today the acceptance of the norm of negative freedom is by no means self-evident. In many countries of the world negative freedom is still rejected. Even in the countries in which it is respected, I think that its effectiveness should be promoted much further, since it increases the probability of leading a good life. Negative freedom enables each of us to conceive our own lives according to our own ideas of the good (Sorgner 2010b, 239–64).

Negative freedom is esteemed as a key norm by all transhumanists who are to be taken seriously as scholars. The significance of freedom, however, differs significantly within different philosophies. Is freedom the only central norm or should equality and solidarity also play an important role? Transhumanism does not give a uniform answer to this question. There are libertarian as well as social-democratic transhumanists. There are also different answers to the question of the philosophical status of the norm "freedom." Is freedom an unchanging moral fact or is it a fragile norm that had to be fought for? In my monograph "Human dignity after/according to Nietzsche," I addressed in detail the hermeneutic argument that the norm of freedom is a norm achieved within the framework of the Enlightenment, the effectiveness of which we must constantly fight for anew (Sorgner 2010b).

6. HUMAN SELF-OVERCOMING AND THE GOOD LIFE

Human beings turn to a variety of technologies in pursuit of their own ideas of the good life. It is safe to say that constant self-overcoming often results in an enhanced quality of life. This consideration is central to transhumanist thought and thus bears additional emphasis here. Children enjoy a period of minimal worry whenever they have the luxury of growing up protected in their parents' house. At that age our cognitive abilities, intelligence, and knowledge have not yet reached their potential. As adults, by contrast, our lives are marked by worry, fear, and difficulty. Nevertheless, many adults would not want to be children again even if that were possible. Despite the pains of adulthood, adults are aware of the abilities they have gained and feel good about possessing them. Cognitive abilities, intelligence, and knowledge are assets with intrinsic value for many people. Just as we could not imagine what possessing these assets would be like when we were children, we are at present unable to imagine what it would be like to be a posthuman, a being whose qualities are equally beyond us. In light of this analogy, we can proceed on the premise that being posthuman may be a condition worth striving for (Sorgner 2012a). All transhumanists regard self-overcoming as a worthy enterprise. Evaluating which qualities are desirable remains a matter of intensive research.

7. RENAISSANCE IDEALS OR THE RADICAL PLURALITY OF THE GOOD?

The next relevant question is this: Which characteristics actually promote a better quality of life? Transhumanists are notably divided on this point. A transhumanist like Nick Bostrom defends the validity of the Renaissance ideal of a well-rounded personality with outstanding physical and mental abilities. The bioliberal Julian Savulescu, beloved among transhumanists, holds a commonsense ideal that places value on healthspan, cognitive abilities, and intelligence. I, on the other hand, believe in a radical plurality of the good. Any attempt at a nonformal determination of the good seems to me to be condemned to failure since individual physiopsychological requirements differ too greatly to allow for any nonformal, universally valid description of the good life. However, the considerations that form the basis for my advocating this ideal are not purely philosophical but must also include attention to its social aspects. Even evolutionary considerations do not definitively establish which qualities will be in our interest in the future. Deafness could even be advantageous for humanity under certain conditions, for example, if unforeseen technological or natural processes caused such mind-numbing noise that no earplugs could block it (Sorgner 2016a).

There is a well-known case where this example is relevant. A deaf lesbian couple wanted a child. They did not consider deafness a handicap but merely a form of otherness. Since they also lived in an environment where there were many deaf people, they thought that deafness may even be advantageous in this context. For these reasons, they intended to increase the likelihood of a deaf child by choosing a deaf sperm donor with whom they have also been friends for a long time. Should they be allowed to increase the probability of their having a deaf child in this way? One US state agreed to allow the couple to follow through with their plan. I for one consider the legality of this procedure fully morally justifiable. Here in particular the choice was clearly not made in order to cause damage to the child. Finally, the variations in question are not between child X with the ability to hear and child X who is deaf. Rather, the child is either child X, born with the characteristic of deafness, or child Y, who possesses the ability to hear. I am well aware that many people will react differently to this case. After careful philosophical consideration, however, I consider the basic attitude presented here to accord with the values of a pluralistic state. For Savulescu, by the way, this moral question decisively led him to revise the concept of good that he had previously advocated. In 2001 he felt that disability always hampered the achievement of a good life. In 2009, he revised this view by emphasizing that disabilities are problematic only if they pose a problem in one's own social environment. On the basis of Bostrom's reflections about the question of the good, the wishes and reflections of both mothers cannot be regarded as morally legitimate. This example clearly illustrates the variety of views held among transhumanists.

8. MORAL BIOENHANCEMENT AND/OR THE CORRELATION BETWEEN COGNITIVE AND MORAL DEVELOPMENT?

In my previous comments on deafness, I did not intend to suggest that cognitive abilities are not of particular relevance. I am well aware that they are a concern for many people. It may even be the case that the development of cognitive abilities would raise the probability of moral evolution. Cultural history suggests a correlation between the development of morality and cognitive abilities. Our technical and cognitive abilities have changed radically in the past, and the same applies to the development of morality, which I see as directly linked to the norms of freedom, equality, and solidarity. The relevance and widespread acceptance of these norms have been established only in the past one hundred years. There are solid empirical studies that speak in favor of pursuing our moral and cognitive development, some even suggesting a causal relationship between these development processes.

In his monograph *The Better Angels of Our Nature*, psychologist Steven Pinker has compiled a large number of empirical examples that indicate our moral progress. His historical examples reveal processes that "orient us away from violence and toward cooperation and altruism and that this process has been going on for millennia" (Pinker 2011, xxv). In particular, he provides a detailed examination of our capacities for empathy and reason in the context of these processes, and he counts these two qualities among our "better angels." The book also addresses the meaning of "the concept of human rights—civil rights, women's rights, children's rights, gay rights, and animal rights" (Pinker 2011, xxiv–xxv). The question of cognitive development, on the other hand, was impressively examined by James R. Flynn, who took extensive empirical data into account. He specifically looked to standardized intelligence tests in order to glean solid data from which reliable conclusions can be drawn. He found that our cognitive abilities have improved significantly, at least during the twentieth century. This process is known as the Flynn Effect. Pinker has integrated Flynn's research into his own work and addressed the possibility of a "moral Flynn effect."[5]

The thought that cognitive development could be coupled with moral development was already discussed centuries earlier by Kant in "Anthropology from a Pragmatic Point of View." Kant's philosophy can also convincingly illustrate the connection between moral and cogitative qualities, since both abilities are, for him, bound up with reason. If there is indeed a causal connection between moral and cognitive developments, and the data just mentioned suggest this insight, then the promotion of cognitive enhancement could also indirectly enhance morality. In any case, Kant's version of the argument seems more promising than the concept of direct moral enhancement as proposed by Persson and Savulescu (2012), among others. Their demands arose from the consideration that before cognitive enhancement can take place, morality must first be enhanced in order to minimize the risk of the extermination of all human beings through the implementation of new technologies. Such an outcome would be probable only if moral bioenhancement is introduced globally in a way that made it compulsory, a proposition I do not consider realistic. Besides all of that, I do not see reliable moral bioenhancement technology coming any time soon. Transhumanist proposals in which moral bioenhancement plays a central role thus strike me as implausible. If my own considerations, briefly summarized here, are plausible, then enhancing cognitive abilities would indirectly lead to enhanced moral behavior (Sorgner 2016b). In any case, it should be noted that moral development in the past centuries has led to the emergence of an enlightened culture in which the norms of freedom, equality, and solidarity occupy a particularly elevated position.

9. NEW IMAGE OF THE FAMILY

The norm of freedom strikes me as particularly important among the progressive norms that have achieved widespread acceptance. This norm also promotes the pluralization process with regard to various concepts, including that of the family. In February 2015, a new biotechnology became legally permissible in Great Britain making it possible to produce children with three biological parents so that mothers with a genetically transmissible mitochondrial defect can have their own children without this defect. This capability is currently unique in the world. The question of triple biological parenthood arose on the basis of successful experiments with great apes. One biotechnologically feasible scenario is that one egg cell is enucleated and the nucleus of another mother is inserted. Since there is also genetic material in the enucleated cytoplasm[6] of the original egg, this procedure results in an egg with genetic material from two mothers that can be fertilized by a sperm cell, thereby creating a child with three biological parents.

This technology seems to offer an appealing option, not only for mothers with genetic defects to have healthy children, but also for lesbian couples to have their own biological children. Triple parenthood makes it possible for a family of two women and a man to have children with genes from all three parents. If the adults concerned love each other and want to live together as a family, then this wish seems completely legitimate to me. It certainly seems questionable that a state should forbid people who love each other from living together in solidarity as a family. Three biological parents should be allowed also to marry each other if they wish. Many other transhumanists are also in favor of expanding the concept of the traditional family, for example, James Hughes and Martine Rothblatt.

Philosophically, transhumanism goes hand in hand with a turn away from natural law. Although transhumanists usually uphold a form of naturalism, the transhumanist concept of "nature" differs from that of natural law. The word "nature" sounds too precious, and it can be identified with many possible meanings, usually requiring more precise clarification. In natural law, the term "nature" refers to a higher nature that is not empirically accessible but rather abides as an eternal essence. This framework presumes that every entity contains an inherent purpose, and that genitals exist for nothing other than reproduction. The family ideal corresponding to this view can contain only one man, one woman, and their biological children. Nature, as discussed by transhumanists, has no such implications but often stands for everything that is empirically accessible to us. It does not, however, make any demands about what should happen (*Sollensforderungen*). Is (*Sein*) and should (*Sollen*) are clearly separated from each other. To say what is means to represent the world descriptively. To say what should be requires us to make statements that encompass norms and values that are human-made and whose effectiveness depends on power relations. The newest biotechnologies make it more possible than ever for us to renounce the conditions of traditional reproduction. Why should we limit the reproductive techniques available to us? Genetic diversification and transformation are always important for the further development of life forms since survival itself depends on adaptation, that is, on how well an organism is adapted to its own environment. For its part, the environment is also subject to a process of constant change and becoming, which is why it is unclear which characteristics will be appropriate in the future. Technology can already promote greater diversification of the means of human reproduction. These considerations suggest a departure from traditional natural law, so that a transhumanist "Natural Law 3.0" can be developed, which would bring with it the possibility of marriage and family formation by three biological parents. However, it is not necessary to refer to any concept of nature to approve of this liberal policy. It is sufficient to hold negative freedom as a desirable achievement. A liberal-democratic state that takes its own liberality seriously should also consider this form of family to be legally legitimate.

10. BEYOND SPECIESISM

A further question arises when we consider the possibilities of human hybridization. Research in this field is already underway at several universities in Great Britain, where the organisms developed for this purpose must be destroyed within two weeks of their creation. Which moral criteria speak against the further development of human-animal hybrids? Human dignity cannot be cited in this respect, since the category of the human cannot be applied to parahuman beings. This research could well lead to a great many new possibilities.

Here is one consideration to keep in mind: Dutch researchers have shown that zebra fish can be genetically modified in such a way that they are able to generate energy by photosynthesis, supplementing the energy they get from their diet (de Lange, n.d.). The fish also take on a light green coloration. Such modification or hybridization with plants could also be interesting for humans facing the problems of population growth and limited resources. Perhaps the little green men from Mars out of science fiction novels are in fact our own descendants.

The works of transgender transhumanist Martine Rothblatt provide another example. One of her projects involves the genetic modification of pigs. She conducts research in this area on her own pig farm with the goal of making it possible to extract organs for xenotransplantations that will not be rejected by a human organism. Her motivation partly derives from a personal misfortune: her daughter suffers from a life-threatening lung disease. Rothblatt hopes that such a treatment may someday be available for her daughter.

To go beyond the boundaries of speciesist thinking is a central concern for transhumanist morality and is closely related to a fundamentally naturalist way of thinking. The aforementioned embrace of nonhuman personhood is also an expression of this kind of thinking. Nonhuman personhood and the overcoming of speciesism, however, affect not only the human-animal relationship but also the human-machine relationship.

A vivid example here is the fictitious android Data from the television series *Star Trek: The Next Generation*. Data's superintelligence differs categorically from human intelligence. Superintelligence may also be accompanied by increased autonomy, a kind of hyperautonomy. If the moral status of living beings should be linked to morally relevant qualities and no longer to their membership in the human species, then Data should be granted rights. Should these rights go beyond those of humans or not? If autonomy is a morally relevant characteristic and Data categorically differs from humans in this respect, then it would be obvious that Data should be granted a higher moral status, for example, a postperson status. Analogous arguments could also be made regarding other morally relevant characteristics, such as those of beings with a meta-self-awareness or a capacity for nanosensitivity. At some point it may also be possible to load human consciousness onto a hard disk, making the cyborg a transitional form toward a new species. The radical nature of the issues involved can hardly be underestimated. It is also an open question in whose interest such a development would be. Would it have the consequence that, in the future, physical existence could become a form of existence for the superrich, or would continued existence as digital posthumans with an enormously increased life span be the more desired mode of being? In any case, we are confronted with numerous delicate challenges, if we seriously aim to transcend speciesism.

11. "BRAIN PACEMAKERS FOR ALL!"

Meanwhile, the limits of our humanity are being blown apart by the arrival of cyborgization. Deep brain stimulation has proven particularly successful in treating Parkinson's disease and severe depression. Research by the engineer Kevin Warwick clearly shows the enormous potential of human-machine interfaces. He has managed to connect neurons from a rat brain with mechanical apparatuses so that they can develop separately, which actually strengthens their neuronal networking (Warwick, n.d.). Even if we cannot yet speak to the feasibility of mind uploading, the potential of brain pacemakers can hardly be underestimated. It is probably only a question of time until Apple presents the *iBrain* and Sony the *Brainstation*.

Generally, cyborgization and research in the field of artificial intelligence are much more dominant in transhumanist discourse than their depiction in this overview would suggest. This has to do with the fact that, as I see it, the further development of humanity by means of genetic engineering will have a greater impact in the foreseeable future than cyborg techniques or AI. I don't want to say that the latest developments in AI and robotics are progressing slowly. Such a claim is certainly unwarranted. At least one transhumanist, Bostrom, disagrees on this topic, which is why his 2014 monograph is focused on "superintelligence." His treatise led to an intense international public debate, in which even Bill Gates participated.

Andy Clark and David Chalmers's Extended Mind Theory deserves attention here. The theory implies that the mind need not be limited to the area within the skull but that objects outside the body can also function as parts of the human mind. Spirit-machine interfaces can be either invasive or noninvasive. A direct local coupling is not even absolutely necessary given the current state of internet technology. This powerful and exciting theory makes it clear that cyborgization has already made significant progress. It can also have practical implications. In Germany, no parts of the bodies of living people are considered mere objects, as long as they are not separated. The first tantalizing question here is how the concept of separation plays out. Do sensors located at a distant location count as unseparated body parts when they are connected to a brain over the internet? Should they count as such? They would, after all, make it possible to experience remote tactile sensations. Moreover, they are connected to the brain even though the connection is not physiological but rather mediated by the internet. Does this make a difference? It makes no meaningful difference with respect to the sensory experience. In both cases the person concerned is capable of conscious tactile experience.

This example is exciting because the technologically mediated (*technisch*) body part is physically distant from the person to whom it belongs. However, there are also numerous cases in which there is no such local separation. A cochlear implant is an invasive human-machine prosthesis for the deaf that enables them to hear. An adjustment to the implant can be made using a smartphone with a wireless connection. The question arises as to which entities can be considered unseparated body parts. Does the implant count? What about the smartphone that regulates its settings? Both views could be considered plausible. However, to regard the smartphone as an unseparated part of the body would also mean that the confiscation of a smartphone from such a cyborg could be construed as bodily harm.

12. METAHUMANITIES AND THE FUTURE OF EDUCATION

The various developments described here do not present us merely with important ethical and anthropological challenges. They also suggest crucial revisions of moral criteria and ontological assessments, especially in the context of education, though this question has not yet received decisive attention from a transhumanist perspective.[7] In 2014 I was invited to address this topic at the UNESCO World Humanities Forum (in Daejeon). The twelfth pillar of transhumanism outlined in this chapter is based on the thoughts I presented there.

A central consideration here is the concept of the humanities. The humanities in a broad sense have taken shape under the influence of the dualisms of Varro, Cicero, and others since their beginnings, as I have already explained. Such an anthropology, however, seems implausible against the background of current developments such as epigenetics and embodied theories of mind. This theory of the human should be revised so that the metahumanities can emerge. University-level metahumanities could begin with a new approach to education that begins with new forms of parental education. The following three points are of particular importance, with the first two points for parental education and the third point for the academic humanities (or metahumanities).[8]

(1) *Genetic enhancement and education as structurally analogous processes*: Traditional education could be facilitated by the modification of genes by parents, as long as they operated within an established framework. I have discussed this point in more detail elsewhere.

(2) *Genetic analysis as a supplemental prerequisite for the educational process*: In order to better recognize and accurately classify the abilities of a particular child, a careful genetic analysis could be of great benefit both in the educational context and with regard to health care. This consideration holds enormous potential and contains consequences that are of incalculable importance for our world, for example, the question of bioprivacy, the organization of our insurance system, and prerequisites for employment. At this point, I can address only a few points to hint at the sensitivity and relevance of this issue. In the United States, Ireland, and Iceland in particular, there is a widespread willingness to make one's own DNA available. Such a vast amount of data can help determine correlations between the presence of certain genes and other characteristics (diseases, abilities, etc.). However, it must be kept in mind that statements made on the basis of genetic material are always associated with certain probabilities. One's own interest or a parental interest in such information can be enormous, since with the help of this data it is possible to take appropriate preventive action, to address weaknesses, and to assess dangers in order to prepare for them in the best possible way. However, health insurers and potential employers also have an enormous interest in this data. This is where a question arises that can be expected to become increasingly relevant in the coming years: bioprivacy. Who is entitled to what information and how are infringements sanctioned? It should also be noted here that it cannot be ruled out that genetic information may be passed on to a potential employer, even if employers are not willing to have access to this

information. We should also bear in mind here that one shares a large part of one's genes with one's own siblings. The challenges, however, go far beyond this question. Could a health insurance company be punished if it obtains this information without permission, or would that be treated as a minor violation? Could it not become obligatory for all people to pass on biometric information to health insurers from birth or to the government, as was legally obligatory in Kuwait from 2015 until 2017? Once a genetic analysis is available and digitally stored, is it not inevitably already public, since everything digital in this day and age could perhaps already be regarded as public? Does it make a difference for the acquisition of certain insurance policies whether or not one has already carried out a genetic analysis? Section 18 of Germany's Genetic Diagnostics Act sheds an interesting light on this topic:

§18 Genetic investigations and analyses in connection with the conclusion of an insurance contract

(1) The insurer may not, neither before nor after conclusion of the insurance contract
1. require genetic tests or analyses to be carried out; or
2. communicate the results or data from genetic tests or analyses already carried out, or accept or use such results or data. For life insurance, occupational disability insurance, disability insurance, and long-term care insurance, sentence 1 no. 2 does not apply if a benefit of more than 300,000 euros or more than 30,000 euros annual pension is stipulated.
(2) Previous and current illnesses are to be indicated; to this extent §§19 to 22 and 47 of the insurance contract law are to be applied (Bundesrecht, n.d.).

This means that anyone who is considering taking out an insurance policy with significant insurance benefits should seriously consider not undergoing genetic analysis. However, as indicated earlier, it could be of enormous value for individuals to obtain this information. These brief reflections alone should make it clear that gene analyses in connection with the presence of big gene data will represent one of the central challenges of the future. Quick fixes for these challenges are not yet foreseeable. Yet it can hardly be doubted that gene analyses will be of immense relevance.

(3) *Metahumanities*: Unlike the humanities, the metahumanities bring nondualistic insights into the classroom, such as the dissolution of the gene-environment distinction in epigenetics, bioart, evolution-ary epistemology, aesthetics, ethics, or economy; embodied theories of mind; new kinds of spirituality and mysticism; nondualistic concepts of rights and dignity; revised family understandings; immanent notions of the good life; the significance of cultural history for norms and values; nonanthropocentric ontologies; and the effort to avoid speciesism. This proposal does not demand that dualistic thinking be banned from the classroom. Instead, it emphasizes the value of integrating nondualist thinking into teaching and of supplementing existing curricula with this dimension. The term metahumanities can also be understood as a further development of the humanities through the inclusion of post-human studies. Since 2017, Penn State University Press has published the *Journal of Posthuman Studies*, which is further advancing academic work on such topics. In the following I will describe a specific example from media history: the birth and death of dualistic media.

The birth of the ancient drama in the sixth century BCE was the moment when various forms of dualism explicitly emerged. With the process of institutionalization came spatially dualistic dividing structures, like those developed for theater architecture. Before the institutionalization of the theater, chants took place in the context of Dionysian festivals where singers, dancers, and fellow celebrants were not spatially separated from each other but rather sang, drank, and celebrated as one large group. Through the gradual process of institution-alization, a theater building with a stage and an auditorium developed, leading to a rigid dualistic architectural separation between actors and audience. Other dualisms emerged as part of the same process.

The separation between actors and the public notably meant that spectators found themselves in a half-circle built upward around the stage so that the actors could go about their duties undisturbed. A fur-ther separation kept the chorus and the protagonists apart. On one side, a chorus emerged, whose task it was to sing and dance. On the other side, actors recited from their scripts. The architecture of ancient theaters served to reinforce the effects of dualities even further, since they were structured in such a way that categorical dualisms could be consolidated and expressed visually. These dualisms were unknown before the institutionalization of the theater, but they resulted from the creation of the Theater of Dionysus in Athens in the sixth century BCE (MacDonald and Walton 2011).

An analogous process occurred around the same time in philosophy when Plato separated the material world clearly and rigidly from the world of ideas. The Stoics extended this innovation by claiming that a reasonable soul (*Vernunftseele*) inhered in all humans and that this insight had ethical consequences. A third decisive step occurred when Descartes asserted that plants and animals were machines, that is, purely material entities without souls, a view that strongly contributed to the promotion of substance dualism. This way of thinking can be found also in the philosophy of Immanuel Kant. Only with Darwin and Nietzsche were the first major nondualistic thinkers within the European canon. That nondualistic turn is now well represented in a variety of contemporary media. With a quick look at the work of Kevin Warwick, Dale Herigstad, Jaime del Val, and Sven Helbig, I will sketch four different variants of nondualistic media. Kevin Warwick's works are particularly closely tied to transhumanism, Dale Herigstad's media creations illustrate posthumanist considerations, and Jaime del Val's metaformances and Sven Helbig's compositions represent key metahumanist positions.

Kevin Warwick

Kevin Warwick's works clearly show the potential of the latest technologies. They also make a case for the development of these technologies and point to a nondualistic view of humanity. All three of these points indicate the presence of transhumanist ideas. One of his well-known innovations makes these facets of his thinking particularly clear. During a stay at Columbia University in New York, he used a brain-computer interface, which he had developed himself, to send his thoughts over the internet to a computer in his laboratory at the University of Reading, which was in turn connected to a mechanical arm. Simply by thinking, he sent out signals that steered the mechanical arm with the hand on the end of it in the direction of a table. The hand grabbed the table, and pressure sensors on it detected the surface of the table and sent the signals to Warwick's brain over an internet connection. He was able to consciously perceive the physical contact with the table while sitting in New York (Tharwat 2012). No one before him dared perform such an experiment, an experiment he performed on himself even before it was tested on apes. There was a risk that the feedback mechanism could have destroyed the structure of his brain. However, he was able to demonstrate the possibilities of this type of technology and make it clear that the categorical distinctions between mind and body and between inorganic and organic substances should be softened.

Dale Herigstad

Digital media artist Dale Herigstad was responsible for the media design of the Steven Spielberg film *Minority Report* and has won four Emmy Awards. Recently, he developed a technology that allows him to use a smartphone app to surf virtually in a three-dimensional space while watching his 3D television set and at the same time change the spatial relationship between the television picture and internet data (TEDx Talks 2013). In the future, this technology will be operated with gesture control, whereby a sensor will analyze the user's muscle tensions in order to perceive the corresponding signals and integrate them into control commands. In the next step, the 3D screen will be replaced by 3D glasses, which in turn will be replaced by a contact lens or an eye implant. In this way, the perceiver is placed within the three-dimensional digital space, abolishing the traditional dualistic distinction between the perceiver and the mediated perception.

Jaime del Val

Jaime del Val, longtime director of the EU-funded Metabody Project, has appeared both as an artist and as a public metahumanist intellectual. A great example of his artistic interrogation of various dualisms is an older metaformance of his entitled "Pangender Cyborg" (del Val 2010). Within this metaformance, Jaime del Val moves through a public space wearing a self-developed apparatus consisting of a video projector, loudspeakers, several cameras, and a microphone. The cameras are directed at his otherwise naked body from unusual perspectives. A postanatomic perspective is always projected onto the wall in front of him by means of the projector. He emits sounds that express spontaneous responses to his spontaneous movements. Various philosophical theses are conveyed in this way: (1) a relational ontology that weakens the duality of subject and object, (2) the representation of postanatomic perspectives reveals the contingent reification of the traditional anatomy, (3) the plurality of sexual orientations or relational metasex is proposed here as an option.

During this metaformance, Jaime beholds his own body from whichever perspective is projected onto the wall in front of him. This perspective is enlarged when displayed on the wall, and he has to move slowly since the camera is sensitive to small movements at that range. Since everything that surrounds a person has an effect on the person, this is just as much the case in the relationship between the organs of perception and the organs perceived as images. This project is one way to illustrate relational being-in-the-world. The categorical subject-object split is thereby blurred.

The significance of this thinking is hardly to be underestimated. The dissolution of the subject-object distinction here goes hand in hand with the image of an amorphous body, a metabody. The very possibility of observing a body in this way undermines the subject-object distinction. To collapse the difference between body and image leads to a collapse of the difference between interior and exterior. This kind of monism could have effects on the ethical evaluation of xenotransplantations, that is, the transfer of functional cells between members of different species. In order to illustrate this nondualistic perspective, it is sufficient to take the number of bacteria in the human body into consideration. About 100 billion microorganisms are part of every single human body, and human life is unthinkable without them. This is especially relevant now that the question of the evaluation of human-animal hybrids is under consideration. (Recall that such hybrids were legally created in England, as long as they were killed again after fourteen days.) Jaime's metaformance reminds the audience that the human being is already a hybrid being in some regards.

But Jaime del Val's metaformance also highlights other points. The projections on the walls also question the traditional description of anatomy; the postanatomical perspectives of the cameras highlight the contingency of the categories present, showing that the anatomy of the body could be mapped quite differently. If we want to, we can depart from calcified structures. Such new structures could lead to the rejection of the prohibition of incest in adults.[9] Why should adults not have the right to mate in agreements that they freely choose? The same insight can be applied to other areas. It is not absolutely necessary to control the computer with a mouse or by swiping on a tablet surface. If the muscle tensions of the arm can be appropriately tapped, gesture control alone may also be an option someday. Paradigm shifts typically require us to part ways with old, calcified structures so that we can reinterpret them in innovative ways.

Another main thrust of Jaime del Val's metaformance is the reconceptualization of sexuality. It questions the traditional gender duality and makes it clear that erotic relations and affects can be linked to more than just the primary male and female gender characteristics. A man does not love every woman just because she has female characteristics. What fascinates and attracts us can be something as simple as someone moving a thigh muscle, raising an eyebrow, or holding a certain posture. Metasex is the pluralization of sexual relations and the associated dissolution of a binary sexuality. In Jaime del Val's "Pangender Cyborg" metaformance, we are confronted with an apt illustration of the pluralization of the good, which also implies a pluralization of the family model.

Sven Helbig

Sven Helbig is, in my estimation, the most important living German composer. His oeuvre also contains numerous structural analogies with metahumanism. His works for the stage can be assigned to the tradition of the *Gesamtkunstwerk*. Moreover, some anticipations of posthuman positions can be found also in Richard Wagner, who is Helbig's most important forebear in this respect. For example, it could be pointed out that the gods of Wagner's opera *Rheingold* are dependent on eating Freia's golden apples in order to retain their divine qualities of strength and youth. There are therefore structural analogies between these gods and the posthumans, as described by transhumanists, for whom posthumans represent an evolved form of human being. However, there are also references to posthumanist positions in Wagner's work. The language used in his musical dramas should be mentioned here, for example. Not only does this sound unusual for us today, but it also does not represent the everyday language of the nineteenth century. Wagner was aware that ideological content is always conveyed with words and that a rigid subject-object distinction is closely linked to the German language. German itself undergirded immanent, naturalistic, and evolutionary thinking. In order to avoid the dualist implications of German grammar, he developed his own personal, metaphorical language, which he used within his musical dramas. These two examples show that both post- and transhumanist elements can already be found in Wagner's works, which is why he can also be seen as a forebear of metahumanism, since this line of thought is located between trans- and posthumanism. The general social, political, and ethical orientation of his work, however, contains a basic attitude that is in conflict with metahumanist thinking and that is connected to numerous potentially problematic implications. This assessment does not apply to Helbig's work. His musical drama *Vom Lärm der Welt oder die Offenbarung des Thomas Müntzer* successfully avoids the potentially totalitarian connotations of *Gesamtkunstwerken*. Nevertheless, it addresses metaphysical, ethical, and political questions. It does not, however, remain isolated within the mythical realm but always refers to current bioethical and religious challenges. What should the relationship between religious and political attitudes be? Which moral assessment is appropriate with regard to the ethical questions at the beginning of life? Should utopias play a role in everyday political decisions? In the final scene of *Vom Lärm der Welt*, the demons emphasize that we are doomed if we adhere to utopia, the established order, and strong beliefs. In this way, the piece emphasizes the radical plurality of the good and the fact that the ethical nihilism of our time is an achievement and not a loss. These are metahumanistic insights. Using the latest technologies, innovative media and an accessible musical language, Helbig avoids limiting the reception of this work to a specialized audience.

Analogous considerations must also be made with regard to his concert *Pocket Symphonies Electronica*. In this work, he draws on orchestral recordings of his own music, plays along live as an instrumentalist, and at the same time is responsible for the appropriate mix, thus also taking on the role of DJ. This undermines the separation between live and recorded music, between serious and popular music, and between the composer and the musician, a difference that has also existed since ancient theater. His instrumental musical work is thus an inspiring plea for plurality and the dissolution of rigid traditional categories.

CONCLUSION

On the foundation of the twelve pillars mentioned here, I have briefly outlined the weak variant of transhumanism that I defend. I have also provided an overview of the diversity of transhumanist positions within a multitude of philosophical discourses, making it clear that other transhumanists adhere to much stronger variants of transhumanism than my own. In particular, hopes for mind uploading, cryonics, and super intelligence are particularly widespread among many of the world's leading transhumanists. Often there is even talk of a moral duty that should exist with regard to the application of certain techniques, for example, Savulescu's case for the moral duty to select certain fertilized ova after artificial insemination and the subsequent preimplantation diagnosis (Sorgner 2014a). I am rather critical of these hopes and objectives, which is why I advocate a weak variant of transhumanism. I think, however, that the considerations proposed here should be taken into account and legally implemented in a liberal, democratic, and pluralistic society. In any case, the considerations and developments mentioned represent a break with long prevailing cultural structures, which are still strongly anchored but have lost their plausibility to a large degree. The time has come to promote a more widespread institutional recognition of today's newly emergent cultural ideals.

Concluding Thoughts

The previous remarks do not claim to deal comprehensively with the subject of transhumanism. A central concern, however, was to convey why transhumanism is for many the most dangerous idea in the world, an assessment that should make sense to anyone who is not sympathetic with naturalism and the theory of evolution. Transhumanism in all its facets is indeed a challenge to traditional dualistic thinking. It represents a paradigm shift that was decisively initiated by Darwin and Nietzsche and whose effectiveness becomes clear when we consider the emergence of numerous new scientific disciplines where evolutionary theory plays a central role, including evolutionary biology, evolutionary aesthetics, evolutionary economics, evolutionary game theory, evolutionary economic geography, and evolutionary epistemology, among others.

The special task that accompanies transhumanism is the permanent need to take on new challenges since every breakthrough in technological innovation raises new social, economic, ethical, and anthropological questions. In 2013, for example, Carl Frey and Michael Osborne of Oxford University made it clear that 47 percent of all employees in the United States work in an industry with a high probability of automation (Frey and Osborne 2013). The exact consequences of this anticipated shift are still difficult to estimate but of enormous social significance. Will it become cheaper in the future to have human employees, when many areas of activity will be automated, radically increasing the number of unemployed, or will the employment of human workers become a niche product for the superrich who value the human touch? In my view, developments in bioprivacy, big gene data, and the internet panopticon are particularly relevant for the social consequences of new technologies. The concept of the panopticon has an exciting philosophical history from Bentham to Foucault to the present day, where it has developed an unprecedented significance following the emergence of the digital world. I have dealt briefly with this topic when discussing the future of parental education. The challenges associated with it naturally go far beyond the topics noted there. The same applies of course to all of the topics addressed in these reflections. Even without being able to anticipate the internet, Foucault aptly described the processes associated with the internet panopticon: "He who is subjected to a field of visibility and who knows it, assumes responsibility for the constraints of power; he makes them play spontaneously upon himself; he inscribes upon himself the power relation in which he simultaneously plays both roles; he becomes the principle of his own subjection" (Foucault 1995, 202–3). Much could and should still be said here, and, with the rapid progress of digitization processes, the relevance of this consideration is constantly increasing in importance. Escaping the internet panopticon is no longer a realistic option for the citizens of technologically advanced states. We will all necessarily become prisoners of this system. How to deal with this is an enormous challenge that urgently requires further thought. Based on what we search for on the internet, what we write in our emails, and which websites we visit, highly specific psychological personality profiles can be created. Using GPS data gathered by smart phones, navigation devices, and public surveillance cameras with facial recognition, our movement patterns can be reconstructed. Biometric data, genetic analyses, fitness apps, and the digital storage of our medical history provide an enormous spectrum of information about our physiological traits. Personality profiles, physiological characteristics, and movement patterns are available as digital data and are transmitted over the internet. The largest internet hub in the world is called DE-CIX and is located in Frankfurt. Whoever is able to get at all of this information by means of such a node might even accurately predict a person's actions. The possible consequences for numerous areas of our lives are so complex that they cannot yet be estimated.

One technology whose effects remain vastly underestimated is genetic analysis. With big gene data, new correlations between abilities, properties, and genes can always be identified. The personal benefit of this information can be enormous. However, the interest of the state, potential employers, and insurance companies is also great. Once the relevant data is digitally available, it is no longer private. Kuwait enacted a law in 2015 that obliged all residents and visitors to perform a DNA analysis. It was abolished in 2017. The social consequences of such developments can hardly be foreseen, especially since the question of the bioprivacy of others is also in play whenever one shares genetic information. My DNA says something not only about me but also about any possible siblings. If someone has his genome analyzed and has the results published on the internet, it may have serious consequences for those siblings. These challenges urgently need to be taken seriously and discussed publicly.

The preceding reflections do not represent a radical critique of the technological development of the digital world. This process is also partly responsible for the fact that the average life span in Europe, North America, and Australia has exceeded eighty years, whereas in Nigeria, one of the world's poorest countries, it is about fifty years. However, it is intended to illustrate the importance of ongoing debate on the latest technologies. Transhumanism promotes a constant and critical confrontation with what has already emerged, what is in the process of emerging, and what will probably emerge next. Every technology brings its own dangers. The appropriate reaction, however, is not a "back to nature" mentality, but an intensive, transdisciplinary, and global confrontation with the resulting challenges, so that new technology is regulated on the best possible basis with respect to the most relevant norms and values. The thoughts presented here also suggest some possible solutions. I am confident that on the basis of intensive, informed discussions, we will arrive at regulations that will make our lives more fulfilling.

But is transhumanism really the most dangerous idea in the world after all? For anyone who believes in an immortal, immaterial soul, it very much is. But even for the rest of us, transhumanism is not without dangers. This does not mean, however, that it should be rejected on that ground. On the contrary. Even the greatest opportunities come with dangers. We take a risk whenever we leave our comfort zone. But only when we expose ourselves to danger can we grow and experience the fullness of life.

NOTES

1. Ambrosia insists that Peter Thiel, the most prominent transhumanist CEO to speak enthusiastically about this technology, is not in fact a client (Cuthbertson 2018).

2. Sorgner discusses this development in the the chapter "The implanted new human" of his monograph *Schöner Neuer Mensch* (Sorgner 2018).

3. Alcor CEO Max More, for instance, reports that most of their patients are men (Alcor Membership Statistics, n.d.). For some scholars, the gender politics of contemporary futurism recall its masculinist origins (Nelson 2002). Filippo Marinetti's 1909 *Futurist Manifesto* links women with nature—and thus with the inferior and stultifying past.

4. The question comes off like an even snarkier ad hominem attack in German since the name "Sorgner" contains the word *Sorge*—German for "worry."

5. Microbiologist Ben Libberton also expresses concern that the microchip implants in Sweden could be the first step toward the sharing of health information in ways that could be used against the sick (Savage 2018). Sorgner addresses these kinds of concerns in his chapter on the internet-panopticon in his monograph *Übermensch: Plädoyer für einen Nietzscheanischen Transhumanismus* (Sorgner 2019).

6. Raadfest, n.d. Vita-More is also the husband of Max More, CEO of Alcor (discussed above), and coeditor of *The Transhumist Reader* together with her husband (More and Vita-More 2013).

7. The questionable autonomy involved when human genes are modified in order to increase individuals' marketable human capital is raised unforgettably in Boots Riley's 2018 film *Sorry to Bother You*.

8. In 2018, Rothblatt was named the highest earning CEO in the biopharmaceutical industry (Carroll 2018).

9. The author's words on this topic: "Metahumans as metasexual: Metasexuality is a productive state of disorientation of desire that challenges categories of sex-gender identity and sexual orientation. A metabody is not ultimately categorisable in terms of morphological sex or gender but rather is an amorphogenesis of infinite potential sexes: microsexes. It is postqueer: we are beyond the understanding of gender as performative. Metasex not only challenges the dictatorship of anatomical, genital and binary sex, but also the limits of the species and intimacy. Pansexuality, public sex, poliamoria, or voluntary sexwork are means to redefine sexual norms into open fields of relationality, where modalities of affect reconfigure the limits of kinship, family and the community" (Val and Sorgner 2011).

10. I also needed human assistance: Stefan Lorenz Sorgner made himself generously available to discuss the translation process. John Fisher helped me review the translation for errors and infelicities. Korey Garibaldi, Stefanie Heine, and Ann Marie Thornburg offered extremely insightful comments on drafts of the introduction. At the Penn State University Press, Kendra Boileau, Alex Vose, and an anonymous peer reviewer all brought careful reading and kind professionalism to the task of bringing this manuscript to publishable form.

INTRODUCTORY REMARKS

1. Sorgner deals with these issues in greater deal in his monograph *Übermensch: Plädoyer für einen Nietzscheanischen Transhumanismus* (Sorgner 2019).

2. The relevance of transhumanist thought to policy making is ever increasing. It will be earth shattering when the question of geoengineering is discussed seriously.

CHAPTER 1

1. A short version of chapter 1, "Is Transhumanism the Most Dangerous Idea in the World?" appeared in the student magazine of the Friedrich Alexander University of Erlangen/Nuremberg (Sorgner 2012b).

2. See the DVD containing Peter Sloterdijk's talk: *Optimierung des Menschen?* (Munich: Quartino GmbH, 2005).

3. The *Journal of Evolution and Technology* has dedicated several issues to the relationship between Nietzscheanism and transhumanism (vol. 20, no. 1; vol. 21, nos. 1–2), and *The Agonist*, published by the Nietzsche Circle, New York has published a special issue to the topic (vol. 4, no. 2). The Institute for Ethics and Emerging Technologies has published key links on this topic on their website: http://ieet.org/index.php/IEET /more/pellissier20120423. Finally, the edited volume *Nietzsche and Transhumanism: Precursor or Enemy?*, edited by Y. Tuncel (Newcastle upon Tyne: Cambridge Scholars Publishing, 2017), reprints much of the above along with newer work on the topic.

4. J. Huxley, "Knowledge, Morality, and Destiny—The William Alanson White Memorial Lectures, Third Series," *Psychiatry* 14, no. 2 (1951): 127–51.

5. One of my central concerns is to promote this intellectual attitude through insights from the continental European philosophical tradition of posthumanism. From this tradition emerges the metahumanism founded by the Spanish artist Jaime del Val and myself. Thinking beyond both Christian and Kantian humanism, I am building a bridge between continental European posthumanism and Anglo-American transhumanism. (The ancient Greek *meta* can mean both "beyond" and "between.")

CHAPTER 2

1. From a medical-social point of view, we must consider side effects of "savantism," which range in severity and overlap with autism.

2. Discussed in more detail in Sorgner 2019.

3. More reflections on this topic can be found in Sorgner 2018.

4. More reflections on this topic can be found in Sorgner 2018, as well as in Sorgner 2019.

CHAPTER 3

1. Fukuyama 2004, 42–43. This chapter integrates core considerations from an article I previously published in English as "Pedigrees" (Sorgner 2014c).

2. This thematic area has stimulated discussion in many contemporary discourses. This raises the following questions: Is there a categorical distinction between traditional and innovative pharmacologically supported meditation techniques? I believe that radical mindfulness can also be promoted through the support of pharmaceuticals. Is the promotion of human performance a problem for human development or does it simultaneously support a fulfilling form of deceleration? In my opinion, the intensive debate on deceleration is too critical of the technical possibilities. In my view, the use of techniques can even lead to social and individual deceleration. From a social level, it becomes relevant here that many areas of human activity are eliminated through automation. On an individual level, it should be noted that increased performance reduces the speed and intensity of challenges. The twenty-four *Caprices* by Paganini are a greater challenge for a young violin student than for Itzhak Perlman.

3. The political aspect of transhumanism is a central and much discussed one. I'm not going to go into it any further here. An excellent presentation of this topic can be found in James Hughes's 2004 monograph, *Citizen Cyborg*, and it is also interesting that transhumanist political parties have started being founded worldwide since 2014.

4. An overview of the topic can be found in the volume *Post- and Transhumanism: An Introduction*, published by Robert Ranisch and Stefan Lorenz Sorgner (2014). The following articles are particularly pertinent: Ranisch and Sorgner 2014; Pastourmatzi 2014; Tirosh-Samuelson 2014. The following overview focuses specifically on the aspect of evolutionary theory: Sorgner and Grimm 2013.

5. It is only recently that new variants of realist philosophies have sprung up, e.g., speculative realism or new realism. It is important to distinguish between new realism, represented by thinkers such as Maurizio Ferraris, Mario De Caro, Mauricio Beuchot, Markus Gabriel, Umberto Eco, Hilary Putnam, and John Searle, and speculative realism, represented by Quentin Meillassoux, Graham Harman, Iain Hamilton Grant, and Ray Brassier. With both approaches, however, I see the danger of the implication of new paternalistic tendencies, which is why I am critical of them.

6. Ludwig Feuerbach had this insight long before Freud, whose texts, like Nietzsche's, were not unknown to the latter: "Man stands with consciousness on an unconscious ground. . . . He calls his body his body and yet is absolutely strange to it. . . . He's a stranger in his own house, is placed on the top of a dizzying hill—below him an incomprehensible abyss" (Feuerbach 1960–64, 10:306).

7. This methodology goes back to Vattimo's "weak thinking," his *pensiero debole*. This means no longer thinking on the basis of "strong" principles such as truth, but to face the situation of nihilism positively, which is best understand as liberation from self-made patterns of thought.

CHAPTER 4

1. See Habermas 2001, 43. The English Nietzsche scholar Keith Ansell-Pearson already dealt with the question of Nietzsche's relationship with transhumanism in 1997. However, he and I have different interpretations of Nietzsche's overhuman.

2. This paragraph is a basic summary of my view of Nietzsche's metaphysics (Sorgner 2007, 39–76).

3. Whether enhanced human beings have been treated as a mere means to an end by their parents can be disputed. Aren't parents interested in enhancement precisely because they love their children and care about them and therefore regard them as an end in themselves?

4. Bostrom 2001. While Bostrom uses the word "eugenics" here, it is worth noting that he is referring solely to state regulated eugenics. See Sorgner 2006b.

5. "Let us suppose that you were to develop into a being that has posthuman healthspan and posthuman cognitive and emotional capacities" (Bostrom 2009, 111).

6. "We may note, however, that it is unlikely that we could in practice become posthuman other than via recourse to advanced technology" (Bostrom 2009, 133).

7. The following statements represent a summary of my Nietzsche interpretation, which I explain in detail in the monograph *Metaphysics Without Truth* (2007).

CHAPTER 5

1. Zoltan Istvan, who ran for US president under the Transhumanist Party he founded, has also presented a stimulating and readable novel, *The Transhumanist Wager* (2013).

2. Instead of the expression "weak (Nietzschean) transhumanism," I could also use the expression "weak posthumanist transhumanism." Spanish artist Jaime del Val and I have coined the term "metahumanism" for our particular way of overcoming humanism. This represents an alternative to post- and transhumanism and can be interpreted, in my own variant, as a special form of weak transhumanism and as a weak posthumanism (see Sorgner 2014c; as well as del Val and Sorgner 2011).

3. For example, it depends whether the haploid chromosome set of the sperm contains an X or a Y chromosome, so that either male or female growth occurs after fertilization by this chromosome. The selection of suitable sperm can also be understood as genetic enhancement through selection. For example, it is discussed whether "family balancing" is a morally justifiable reason for choosing the sex of one's offspring.

4. See Sorgner 2011, section 4. I have dealt with this topic in detail in another article: Sorgner 2015.

5. In Pinker's words: We can now put together the two big ideas of this section: the pacifying effects of reason, and the Flynn Effect. We have several grounds for supposing that enhanced powers of reason—specifically, the ability to set aside immediate experience, detach oneself from a parochial vantage point, and frame one's ideas in abstract, universal terms—would lead to better moral commitments, including an avoidance of violence. . . . Could there be a moral Flynn Effect, in which an accelerating escalator of reason carried us away from impulses that lead to violence?

The idea is not crazy. The cognitive skill that is most enhanced in the Flynn Effect, abstraction from the concrete particulars of immediate experience, is precisely the skill that must be exercised to take the perspectives of others and expand the circle of moral consideration (Pinker 2011, 656).

6. Cytoplasm is the fluid held within the cell membrane and contains organelles including the cell nucleus. Even after removing the cell nucleus, mitochondrial DNA is still contained there.

7. Posthumanist scholars have already engaged in an intensive discussion of this issue, which led to the demand for posthumanities departments.

8. Translator's note: There is a meaningful pun in this sentence. Instead of just "humanities," Sorgner writes *(Leib-)Geisteswissenschaften*, the "(body-)mind sciences," which expands the current German term for the humanities so that its name refers to the embodied aspect of the human mind.

9. In Spain, incest between consenting adults is not illegal.

111

WORKS CITED

Agar, N. 1998. "Liberal Eugenics." *Public Affairs Quarterly* 12, no. 2: 137–55.

Alcor Cryonics. 2015. "Cryonics Testimonial: Alcor Member Matthew Deutsch." https://www.youtube.com/watch?v=sYYMLGQbh64.

Alcor Membership Statistics. n.d. Alcor Life Extension Foundation. https://alcor.org /AboutAlcor/membershipstats.html.

Alcor Procedures. n.d. Alcor Life Extension Foundation. https://alcor.org/procedures .html.

Alderman, L. 2018. "In Sweden, Cash Is Almost Extinct and People Implant Microchips in Their Hands to Pay for Things." *Financial Post*, November 23, 2018. https:// business.financialpost.com/news/economy/swedens-push-to-get-rid-of-cash -has-some-saying-not-so-fast.

Ansell-Pearson, K. 1997. *Viroid Life: Perspectives on Nietzsche and the Transhuman Condition*. London: Routledge.

Badmington, N., ed. 2000. *Posthumanism*. New York: Palgrave.

Birx, H. J. 2006. "Nietzsche." In *Encyclopedia of Anthropology*, edited by H. J. Birx, 4:1741–45. Thousand Oaks, CA: SAGE.

Bittner, U. 2011. "Neuro-Enhancement der Liebe: Wird die Liebe zu einem mediz-inisch kontrollierbaren Phänomen? Philosophisch-ethische Reflexionen zu den jüngsten neurowissenschaftlichen Forschungsergebnissen." *Zeitschrift für medizinische Ethik* 57:53–61.

Bostrom, N. 2001. *Transhumanist Values*. Version of April 18. http://www.nickbostrom .com/tra/values.html.

———. 2005a. "The Fable of the Dragon Tyrant." *Journal of Medical Ethics* 31, no. 5: 273–77.

———. 2005b. "A History of Transhumanist Thought." *Journal of Evolution and Technology* 14, no. 1: 1–25.

———. 2005c. "Transhumanist Values." *Review of Contemporary Philosophy* 4, no. 1–2. http://www.nickbostrom.com/ethics/values.html.

———. 2009. "Why I Want to Be a Posthuman When I Grow Up." In *Medical Enhancement and Posthumanity*, edited by B. Gordijn and R. Chadwick, 107–36. New York: Springer.

Braidotti, R. 2013. *The Posthuman*. Cambridge: Polity.

Bundesamt für Justiz. n.d. https://www.gesetze-im-internet.de/englisch_stgb/englisch _stgb.html#p0133.

Bundesrecht. n.d. https://www.buzer.de/gesetz/8967/a163156.htm.

Carroll, J. 2018. "Who's the Top Earning CEO in Biopharma? Martine Rothblatt Once Again Hits Top Slot with $37M Pay Package." *Endpoints News*, April 30, 2018.

https://endpts.com/whos-the-top-earning-ceo-in-biopharma-martine-roth
blatt-once-again-hits-top-slot-with-37m-pay-package.

Carville, O. 2019. "The Bloody Tale of Ambrosia, the Startup That Wants to Slow
Aging." *Bloomberg*, February 25, 2019. https://www.bloomberg.com/news
/articles/2019-02-25/the-bloody-tale-of-ambrosia-the-startup-that-wants-to
-slow-aging.

Chadwick, R. 2009. "Therapy, Enhancement and Improvement." In *Medical
Enhancement and Posthumanity*, edited by B. Gordijn and R. Chadwick, 25–38.
New York: Springer.

Clark, L. 2017. "Why Elon Musk's Transhumanism Claims May Not Be That Far-
fetched." *Wired*, February 15, 2017. https://www.wired.co.uk/article/elon-musk
-humans-must-become-cyborgs.

CNBC. 2016. "Alcor Freezes People for a Shot at Immortality | CNBC." https://www
.youtube.com/watch?v=LhLTy6AKAxg.

Crockett, M. J. 2010. "Reply to Harris and Chan: Moral Judgment Is More Than Rational
Deliberation." *Proceedings of the National Academy of Sciences* 107, no. 50: E184.

Crockett, M. J., et al. 2010. "Serotonin Selectively Influences Moral Judgment and
Behavior Through Effects on Harm Aversion." *Proceedings of the National
Academy of Sciences* 107, no. 40: 17433–38.

Cuthbertson, A. 2018. "Billionaire Trump Supporter Peter Thiel Denies Being a
Vampire." *Independent*, November 2, 2018. https://www.independent.co.uk/life
-style/gadgets-and-tech/news/peter-thiel-vampire-donald-trump-life-exten-
sion-blood-transfusion-ambrosia-palantir-a8614061.html.

DAK Gesundheit. 2009. https://www.dak.de/content/filesopen/Gesundheitsreport
_2009.pdf.

DeGrazia, D. 2004. "Prozac, Enhancement, and Self-Creation." In *Prozac as a Way
of Life*, edited by C. Elliott and T. Chambers, 33–47. Chapel Hill: University of
North Carolina Press.

Derrida, J. 1991. *Gesetzeskraft—Der "mystische Grund der Autorität."* Frankfurt am
Main: Suhrkamp.

Devereaux, M. 2009. "Cosmetic Surgery." In *Medical Enhancement and Posthumanity*,
edited by B. Gordijn and R. Chadwick, 159–74. New York: Springer.

Douglas, T. 2011. "Moral Enhancement." In *Enhancing Human Capacities*, edited by
J. Savulescu, R. T. Meulen, and G. Kahane, 467–85. Oxford: Oxford University
Press.

Elliott, C., and T. Chambers, eds. 2004. *Prozac as a Way of Life*. Chapel Hill: University
of North Carolina Press.

Esfandiary, F. M. 1974. "Transhumans-2000." In *Woman in the Year 2000*, edited by M.
Tripp, 291–98. New York: Arbor House.

Eßmann, B., U. Bittner, and D. Baltes. 2011. "Die biotechnische Selbstgestaltung des
Menschen. Neuere Beiträge zur ethischen Debatte über das Enhancement."
Philosophische Rundschau 58:1–21.

Faggella, D. 2018. "The Transhuman Transition—Lotus Eaters vs World Eaters." https://
danfaggella.com/the-transhuman-transition-lotus-eaters-vs-world-eaters.

Feuerbach, L. 1960–64. *Sämtliche Werke*. 2nd ed. Edited by W. Bolin and Fr. Jodl. 13
vols. in 12. Stuttgart: Frommann Verlag.

FM-2030. 1989. *Are You a Transhuman? Monitoring and Stimulating Your Personal Rate
of Growth in a Rapidly Changing World*. New York: Viking Adult.

Foddy, F., and J. Savulescu. 2009. "Ethik der Leistungssteigerung im Sport: Medikamente und Gendoping." In *Neuro-Enhancement: Ethik vor neuen Herausforderungen*, edited by B. Schöne-Seifert, D. Talbot, U. Opolka, and J. S. Ach, 93–114. Paderborn: Mentis.

Foucault, M. 1995. *Discipline and Punish: The Birth of the Prison.* New York: Vintage Books.

Freud, S. 2003 [1920]. *Beyond the Pleasure Principle.* New York: Penguin.

Frey, C., and M. A. Osborne. 2013. *The Future of Employment: How Susceptible Are Jobs to Computerization?* Oxford: University of Oxford.

Fukuyama, F. 2002. *Our Posthuman Future: The Consequences of the Biotechnology Revolution.* London: Picador.

———. 2004. "The World's Most Dangerous Ideas: Transhumanism." *Foreign Policy* 144, September/October: 42–43.

Gordijn, B., and R. Chadwick, eds. 2009. *Medical Enhancement and Posthumanity.* New York: Springer.

Grey, A. de., and M. Rae. 2007. *Ending Aging: The Rejuvenation Breakthroughs That Could Reverse Human Aging in Our Lifetime.* New York: Griffin Press.

Habermas, J. 1990. *The Philosophical Discourse of Modernity: Twelve Lectures.* Cambridge: Polity Press.

———. 2001. *Die Zukunft der menschlichen Natur: Auf dem Weg zu einer liberalen Eugenik?* Frankfurt am Main: Suhrkamp.

———. 2003. *The Future of Human Nature.* Cambridge: Polity.

———. 2014. "Autopoietische Selbsttransformationen der Menschengattung." In *Biologie und Biotechnologie: Diskurse über die Optimierung des Menschen*, edited by J. Habermas et al., 27–37. Wien: Picus Verlag.

Hamer, D. H., et al. 1993. "A Linkage Between DNA Markers on the X Chromosome and Male Sexual Orientation." *Science* 261, no. 5119: 321–27.

Haraway, D. 2016. *Staying with the Trouble: Making Kin in the Chthulucene.* Durham: Duke University Press.

Harris, J., and S. Chan. 2010. "Moral Behavior Is Not What It Seems." *Proceedings of the National Academy of Sciences* 107, no. 50: E183.

Hassan, I. 1977. "Prometheus as Performer: Toward a Posthumanist Culture?" *Georgia Review* 31, no. 4: 830–50.

Hauskeller, M. 2009. *Biotechnologie und die Integrität des Lebens.* Kusterdingen: Graue Edition.

Hawkins, S. 2017. "Differential Translation: A Proposed Strategy for Translating Polysemous Language in German Philosophy." *Translation and Interpreting Studies* 12, no. 1: 116–36.

Hayles, N. K. 1999. *How We Became Posthuman: Virtual Bodies in Cybernetics, Literature, and Informatics.* Chicago: University of Chicago Press.

Hildt, E., and E.-M. Engels, eds. 2009. *Der implantierte Mensch: Therapie und Enhancement im Gehirn.* Freiburg im Breisgau: Alber Verlag.

Hilt, A., I. Jordan, and A. Frewer, eds. 2010. *Endlichkeit, Medizin und Unsterblichkeit: Geschichte—Theorie—Ethik.* Stuttgart: Franz Steiner Verlag.

Horster, D. 1991. *Rorty zur Einführung.* Hamburg: Junius.

Hughes, J. 2004. *Citizen Cyborg: Why Democratic Societies Respond to the Redesigned Human of the Future.* Boulder: Westview Press.

115

———. 2010. "Contradictions from the Enlightenment Roots of Transhumanism." *Journal of Medicine and Philosophy* 35, no. 6: 622–40.

———. 2014. "Politics." In *Post- and Transhumanism: An Introduction*, edited by R. Ranisch and S. L. Sorgner, 133–48. New York: Peter Lang.

Humanity+. n.d. https://humanityplus.org/about.

Hurlemann, R., A. Patin, O. A. Onur, M. X. Cohen, T. Baumgartner, S. Metzler, I. Dziobek, et al. 2010. "Oxytocin Enhances Amygdala-Dependent, Socially Reinforced Learning and Emotional Empathy in Humans." *Journal of Neuroscience* 30, no. 14: 4999–5007.

Huxley, J. 1957. "Transhumanism." In *New Bottles for New Wine*, 13–17. London: Chatto & Windus.

IEET (Institute for Ethics and Emerging Technologies). http://ieet.org/index.php /IEET/more/pellissier20120423.

Jones, R., and E. Tschirner. 2015. *A Frequency Dictionary of German: Core Vocabulary for Learners*. London: Routledge.

Kahneman, D. 2011. *Thinking Fast and Slow*. New York: Farrar, Straus and Giroux.

Kass, L. R. 1997. "The Wisdom of Repugnance." *New Republic* 216, no. 22: 17–26.

Knaup, M. 2017. "Stefan Lorenz Sorgner: Transhumanismus. ,Die gefährlichste Idee der Welt'!?" *Philosophischer Literaturanzeiger* 70, no. 1: 30–35.

Kosoff, M. 2016. "Peter Thiel Wants to Inject Himself with Young People's Blood." *Vanity Fair*, August 1, 2016. https://www.vanityfair.com/news/2016/08/peter -thiel-wants-to-inject-himself-with-young-peoples-blood.

Krämer, F. 2009. "Neuro-Enhancement von Emotionen: Zum Begriff emotionaler Authentizität." In *Neuro-Enhancement: Ethik vor neuen Herausforderungen*, edited by B. Schöne-Seifert, D. Talbot, U. Opolka, and J. S. Ach, 189–217. Mentis: Paderborn.

Kurzweil, R. 2005. *The Singularity Is Near: When Humans Transcend Biology*. New York: Penguin.

Lange, Robin de. n.d. "Solar Powered Future Fish || Lustrum Event Leiden University." http://www.robindelange.com/solar-fish-lustrum-event-leiden-university.

Levine, Y. 2018. *Surveillance Valley: The Secret Military History of the Internet*. New York: Public Affairs.

MacDonald, M., and J. M. Walton, eds. 2011. *The Cambridge Companion to Greek and Roman Theatre*. Cambridge: Cambridge University Press.

Maguire, A. M., F. Simonelli, E. A. Pierce, E. N. Pugh, F. Mingozzi, J. Bennicelli, S. Banfi, et al. 2008. "Safety and Efficacy of Gene Transfer for Leber's Congenital Amaurosis." *New England Journal of Medicine* 358:2240–48.

Marcus, G. 2013. "Hyping Artificial Intelligence, Yet Again." *New Yorker*, December 31, 2013. https://www.newyorker.com/tech/annals-of-technology/hyping-artificial -intelligence-yet-again.

Marquard, O. 1981. *Abschied vom Prinzipiellen: Philosophische Studien*. Stuttgart: Reclam.

———. 1989. *Aesthetica und Anaesthetica: Philosophische Überlegungen*. Paderborn: Ferdinand Schöningh.

———. 1995. *Glück im Unglück: Philosophische Überlegungen*. Munich: Wilhelm Fink Verlag.

———. 2000. *Philosophie des Stattdessen: Studien*. Stuttgart: Reclam.

Meyerowitz, J. 1980. *How Sex Changed*. Cambridge: Harvard University Press.

Mill, J. S. 1859. *On Liberty*. London: John W. Parker and Son.

————. 2006. *Utilitarianism / Der Utilitarismus*. Stuttgart: Reclam.

Moore, G., and T. H. Brobjer, eds. 2004. *Nietzsche and Science*. Aldershot: Ashgate.

More, M. 2010. "The Overhuman in the Transhuman." *Journal of Evolution and Technology* 21, no. 1: 1–4.

————. 2013. "The Philosophy of Transhumanism." In *The Transhumanist Reader: Classical and Contemporary Essays on the Science, Technology, and Philosophy of the Human Future*, edited by M. More and N. Vita-More, 3–17. Chichester: Wiley-Blackwell.

More, M., and N. Vita-More, eds. 2013. *The Transhumanist Reader: Classical and Contemporary Essays on the Science, Technology, and Philosophy of the Human Future*. Chichester: Wiley-Blackwell.

Müller, S., and M. Christen. 2010. "Mögliche Persönlichkeitsveränderungen durch Tiefe Hirnstimulation bei Parkinson-Patienten." *Nervenheilkunde* 29, no. 11: 779–83.

Müller-Jung, J. 2013. "Hirndoping boomt an Universitäten." *Frankfurter Allgemeine Zeitung*, January 31, 2013. https://www.faz.net/aktuell/wissen/medizin -ernaehrung/jeder-fuenfte-student-nimmt-pillen-hirndoping-boomt-an-uni versitaeten-12045969.html.

Nayar, P. 2014. *Posthumanism*. New York: Polity.

Nelson, A. 2002. "Introduction: Future Text." *Social Text* 71, no. 20.2: 1–15.

Nietzsche, F. 1985. *Thus Spoke Zarathustra: A Book for Everyone and No One*. Translated by R. J. Hollingdale. New York: Penguin.

————. 1988. *Sämtliche Werke: Kritische Studienausgabe in 15 Bänden*. 2nd ed. Edited by G. Colli and M. Montinari. 15 volumes. Munich: Deutscher Taschenbuch Verlag.

————. 1989. *Beyond Good and Evil*. Translated by Walter Kaufmann. New York: Vintage.

Nussbaum, M. 1995. "Human Capabilities, Female Human Beings." In *Women, Culture and Development: A Study A Study of Human Capabilities*, edited by M. Nussbaum and J. Glover, 61–104. Oxford: Oxford University Press.

O'Kane, Sean. 2018. "Tesla's On-Site Health Clinic Accused of Undercounting Worker Injuries." *Verge*, November 6, 2018. https://www.theverge.com/2018/11/6 /18064326/tesla-factory-worker-injuries-clinic-fremont.

Pastourmatzi, D. 2014. "Science Fiction." In *Post- and Transhumanism: An Introduction*, edited by R. Ranisch and S. L. Sorgner, 227–40. New York: Peter Lang.

Persson, I., and J. Savulescu. 2011. "Unfit for the Future? Human Nature, Scientific Progress, and the Need for Moral Enhancement." In *Enhancing Human Capacities*, edited by J. Savulescu, R. T. Meulen, and G. Kahane, 486–502. Oxford: Oxford University Press.

————. 2012. *Unfit for the Future: The Need for Moral Enhancement*. Oxford: Oxford University Press.

Petersén, M. 2018. "Thousands of Swedes Are Inserting Microchips into Themselves— Is It Because of Their Welfare State?" *Independent*, June 21, 2018. https://www .independent.co.uk/voices/sweden-microchips-artificial-intelligence-contact less-credit-cards-citizen-science-biology-a8409676.html.

Pinker S. 2011. *The Better Angels of Our Nature: Why Violence Has Declined*. New York: Viking.

Poe, E. 2008 [1838]. *The Narrative of Arthur Gordon Pym of Nantucket, and Related Tales*. Oxford: Oxford University Press.

117

Raadfest. n.d. "Dr. Natasha Vita-More. Executive Director, Humanity+, Inc." https://www.raadfest.com/dr-natasha-vitamore.

Ranisch, R., and S. L. Sorgner, 2014. "Introducing Post- and Transhumanism." In *Post- and Transhumanism: An Introduction*, edited by R. Ranisch and S. L. Sorgner, 7–29. Frankfurt am Main: Peter Lang.

Rentrop, M., R. Müller, and J. Baeuml, eds. 2009. *Klinikleitfaden Psychiatrie und Psychotherapie*. Munich: Elsevier.

Robertson, J. A. 1994. *Children of Choice: Freedom and the New Reproductive Technologies*. Princeton: Princeton University Press.

Robinson, D. 1999. *Nietzsche and Postmodernism*. Cambridge: Icon Books.

Rothblatt, M. 2015. "My Daughter, My Wife, Our Robot, and the Quest for Immortality." https://www.ted.com/talks/martine_rothblatt_my_daughter_my_wife_our _robot_and_the_quest_for_immortality?language=en.

Savage, M. 2018. "Thousands of Swedes Are Inserting Microchips Under Their Skin." https://www.npr.org/2018/10/22/658808705/thousands-of-swedes-are-insert ing-microchips-under-their-skin.

Savulescu, J. 2001. "Procreative Beneficence: Why We Should Select the Best Children." *Bioethics* 15, no. 5–6: 413–26.

Savulescu, J., and N. Bostrom, eds. 2009. *Human Enhancement*. Oxford: Oxford University Press.

Savulescu, J., and G. Kahane. 2009. "The Moral Obligation to Create Children with the Best Chance of the Best Life." *Bioethics* 23, no. 5: 274–90.

Savulescu, J., R. T. Meulen, and G. Kahane, eds. 2011. *Enhancing Human Capacities*. Oxford: Oxford University Press.

Savulescu, J., and A. Sandberg. 2008. "Neuroenhancement of Love and Marriage: The Chemicals Between Us." *Neuroethics* 1:31–44.

Schlaepfer, T. E., and K. Lieb. 2005. "Deep Brain Stimulation for Treatment of Refractory Depression." *Lancet* 366, no. 9495: 1420–22.

Schmidt-Felzmann, H. 2009. "Prozac und das wahre Selbst: Authentizität bei psycho-pharmakologischem Enhancement." In *Neuro-Enhancement: Ethik vor neuen Herausforderungen*, edited by B. Schöne-Seifert, D. Talbot, U. Opolka, and J. S. Ach, 143–58. Mentis: Paderborn.

Schöne-Seifert, B. 2009. "Neuro-Enhancement: Zündstoff für tiefgehende Kontroversen." In *Neuro-Enhancement: Ethik vor neuen Herausforderungen*, edited by B. Schöne-Seifert, D. Talbot, U. Opolka, and J. S. Ach, 347–63. Mentis: Paderborn.

Schöne-Seifert, B., D. Talbot, U. Opolka, and J. S. Ach, eds. 2009. *Neuro-Enhancement: Ethik vor neuen Herausforderungen*. Mentis: Paderborn.

Scott-Heron, G. 1970. "Whitey on the Moon." On *Small Talk at 125th and Lenox*, produced by Bob Thiel. Flying Dutchman FDS 131, 33⅓ rpm.

Siefer, W. 2009. *Das Genie in mir: Warum Talent erlernbar ist*. Frankfurt am Main: Campus.

Singer, P. 2002. *Animal Liberation*. New York: Harper.

———. 2009. "Speciesism and Moral Status." *Metaphilosophy* 40, no. 3–4: 567–81.

———. 2011. *Practical Ethics*. Cambridge: Cambridge University Press.

Sloterdijk, P. 1983. *Kritik der zynischen Vernunft*. Frankfurt am Main: Suhrkamp.

———. 1987. *Kopernikanische Mobilmachung und ptolemäische Abrüstung: Ästhetischer Versuch*. Frankfurt am Main: Suhrkamp.

———. 1988. *Critique of Cynical Reason*. Translated by Michael Eldred. Minneapolis: University of Minnesota Press.

———. 1999. *Regeln für den Menschenpark: Ein Antwortschreiben zu Heideggers Brief über den Humanismus*. Frankfurt am Main: Suhrkamp.

———. 2001. *Nicht gerettet: Versuche nach Heidegger*. Frankfurt am Main: Suhrkamp.

———. 2009. "Rules for the Human Zoo: A Response to the Letter on Humanism." Translated by Mary Varney Rorty. *Environment and Planning D: Society and Space* 27:12–28.

Sorgner, S. L. 2006a. "Einleitung." In *Eugenik und die Zukunft*, edited by S. L. Sorgner, H. J. Birx, and N. Knoepffler, 1–11. Freiburg im Breisgau: Alber Verlag.

———. 2006b. "Facetten der Eugenik." In *Eugenik und die Zukunft*, edited by S. L. Sorgner, H. J. Birx, and N. Knoepffler, 201–9. Freiburg im Breisgau: Alber Verlag.

———. 2007. *Metaphysics Without Truth: On the Importance of Consistency Within Nietzsche's Philosophy*. 2nd exp. ed. Milwaukee: University of Marquette Press.

———. 2009. "Nietzsche, the Overhuman, and Transhumanism." *Journal of Evolution and Technology* 21, no. 1: 29–42.

———. 2010a. "Beyond Humanism: Reflections on Trans- and Posthumanism." *Journal of Evolution and Technology* 21, no. 2: 1–19.

———. 2010b. *Menschenwürde nach Nietzsche: Die Geschichte eines Begriffs*. Darmstadt: WBG.

———. 2011. "Zarathustra 2.0 and Beyond: Further Remarks on the Complex Relationship between Nietzsche and Transhumanism." *Agonist* 4, no. 2: 1–46.

———. 2012a. "Posthumane leben besser: Ist der Transhumanismus die gefährlichste Idee der Welt?" *Aufklärung und Kritik* 44, no. 19: 4, 160–73.

———. 2012b. "'Warum ich ein Posthumaner sein will,' oder: Transhumanismus—Die gefährlichste Idee der Welt?" *FAU Review* 1:20–29.

———. 2013a. "Evolution, Education, and Genetic Enhancement." In *Evolution and the Future: Anthropology, Ethics, Religion*, edited by S. L. Sorgner and B.-R. Jovanovic, 85–100. Frankfurt am Main: Peter Lang.

———. 2013b. "Human Dignity 2.0: Beyond a Rigid Version of Anthropocentrism." *Trans-Humanities* 6, no. 1: 135–59.

———. 2013c. "Kant, Nietzsche and the Moral Prohibition of Treating a Person Solely as a Means." *Agonist* 6, no. 1/2: 1–6.

———. 2014a. "Is There a "Moral Obligation to Create Children with the Best Chance of the Best Life"? *Humana Mente: Journal of Philosophical Studies* 26:199–212.

———. 2014b. "Jenseits einer rigiden Konzeption des Anthropozentrismus." In *Umwertung der Menschenwürde: Kontroversen mit und nach Nietzsche*, edited by B. Vogel, 165–92. Freiburg im Breisgau: Alber Verlag.

———. 2014c. "Pedigrees." In *Post- and Transhumanism: An Introduction*, edited by R. Ranisch and S. L. Sorgner, 29–48. New York: Peter Lang.

———. 2015. "The Future of Education: Genetic Enhancement and Metahumanities." *Journal of Evolution and Technology* 25, no. 1: 31–48.

———. 2016a. "Nietzsche, Transhumanismus und drei Arten der (post)humanen Perfektion." *Nietzsche, Foucault und die Medizin: Philosophische Impulse für die Medizinethik*, edited by O. Friedrich, 245–68. Bielefeld: Transcript.

———. 2016b. "The Stoic Sage 3.0: A Realistic Goal of Moral (Bio)Enhancement Supporters?" *Journal of Evolution and Technology* 26, no. 1: 83–93.

———. 2018. *Schöner Neuer Mensch*. Berlin: Nicolai Verlag.

119

————. 2019. *Übermensch: Plädoyer für einen Nietzscheanischen Transhumanismus.* Berlin: Schwabe Verlag.

Sorgner, S. L., H. J. Birx, and N. Knoepffler, eds. 2006. *Eugenik und die Zukunft.* Freiburg im Breisgau: Alber Verlag.

Sorgner, S. L., and O. Fürbeth. 2010. Introduction to *Music in German Philosophy: An Introduction,* edited by S. L. Sorgner and O. Fürbeth, 1–26. Chicago: University of Chicago Press.

Sorgner, S. L., and N. Grimm. 2013. "Introduction: Evolution Today." In *Evolution and the Future: Anthropology, Ethics, Religion,* edited by S. L. Sorgner and B.-R. Jovanovic, 9–20. Frankfurt am Main: Peter Lang.

Sousa, R. D. 2009. *Die Rationalität des Gefühls.* Frankfurt am Main: Suhrkamp.

Stadt Zürich. n.d. https://www.stadt-zuerich.ch/ssd/de/index/gesundheit_und _praevention/suchtpraevention/publikationen_u_broschueren/grundlagen ---konzepte/neuro-enhancement.html.

Stephan, A. 2004. "Zur Natur künstlicher Gefühle." In *Natur und Theorie der Emotion,* edited by A. Stephan and H. Walter, 309–24. Paderborn: Mentis.

Stephan, A., and H. Walter, eds. 2004. *Natur und Theorie der Emotion.* Paderborn: Mentis.

Stevens, A. 1994. *Jung.* Oxford: Oxford University Press.

Stewart, J. B., M. Goldstein, and J. Silver-Greenberg. 2019. "Jeffrey Epstein Hoped to Seed Human Race with His DNA." *New York Times,* July 31, 2019. https://www .nytimes.com/2019/07/31/business/jeffrey-epstein-eugenics.html.

Stiegler, B. 1998. *Technics and Time: The Fault of Epimetheus.* Stanford: Stanford University Press.

Stix, G. 2010. "Doping für das Gehirn." *Spektrum der Wissenschaft* 1:46–54.

TEDx Talks. 2013. "Discovering New Media Space: Dale Herigstad at TEDxTransmedia 2014." https://www.youtube.com/watch?v=mabgFkt9scg.

Tharwat, R. 2012. "Biorobotics: Making the Best of Man and Machine, Prof. Dr. Kevin Warwick." https://www.youtube.com/watch?v=iAjBm5pjtjc.

Tirosh-Samuelson, H. 2014. "Religion." In *Post- and Transhumanism: An Introduction,* edited by R. Ranisch and S. L. Sorgner, 49–72. New York: Peter Lang.

Tuncel, Y. 2017. *Nietzsche and Transhumanism: Precursor or Enemy?* Newcastle upon Tyne: Cambridge Scholars.

Ulla, M., S. Thobois, J.-J. Lemaire, A. Schmitt, P. Derost, E. Broussolle, P.-M. Llorca, and F. Durif. 2006. "Manic Behaviour Induced by Deep-Brain Stimulation in Parkinson's Disease: Evidence of Substantia Nigra Implication?" *Journal of Neurology, Neurosurgery, and Psychiatry* 77:1363–66.

US Transhumanist Party / Transhuman Party—Official Website. n.d. http://transhu manist-party.org/platform.

Val, J. del. 2010. "CYBORG PANGÉNERO /// PANGENDER CYBORG—Jaime del Val." https:// www.youtube.com/watch?v=fmCaxU73qTo.

Val, J. del, and S. L. Sorgner. 2011. "A Metahumanist Manifesto." *Agonist* 4, no. 2: 1–4.

Vattimo, G. 1988. *The End of Modernity: Nihilism and Hermeneutics in Post-Modern Culture.* Cambridge: Polity Press.

Warwick, K. n.d. https://www.youtube.com/watch?v=1-0eZytv6Qk.

————. 2012. *Artificial Intelligence: The Basics.* London: Routledge.

Wolfe, C. 2010. *What Is Posthumanism?* Minneapolis: University of Minnesota Press.

Zhou, D. 2018. "Janelle Monáe on Afrofuturism." https://vimeo.com/264269855.

INDEX

122

129